Miller's Guide to Home Remodeling

Home Reference Series

Miller's Guide to Home Remodeling

MARK R. MILLER
Associate Professor
Texas A&M University–Kingsville
Kingsville, TX

REX MILLER
Professor Emeritus
State University College at Buffalo
Buffalo, New York

GLENN E. BAKER
Professor Emeritus
Texas A&M University
College Station, TX

McGraw-Hill

New York Chicago San Francisco Lisbon London
Madrid Mexico City Milan New Delhi San Juan
Seoul Singapore Sydney Toronto

The McGraw·Hill Companies

Cataloging-in-Publication Data is on file with the Library of Congress.

1 2 3 4 5 6 7 8 9 0   QPD/QPD   0 1 0 9 8 7 6 5 4

ISBN 0-07-144553-6

The sponsoring editor for this book was Larry Hager, the editing supervisor was Caroline Levine, and the production supervisor was Sherri Souffrance. It was set in ITC Century Light by Wayne Palmer of McGraw-Hill Professional's Hightstown, N.J., composition unit.

McGraw-Hill books are available at special quantity discounts to use as premiums and sales promotions, or for use in corporate training programs. For more information, please write to the Director of Special Sales, McGraw-Hill Professional, Two Penn Plaza, New York, NY 10121. Or contact your local bookstore.

Contents

Preface

Miller's Guide to Home Remodeling is written for anyone who needs to know about this part of the carpentry and construction field. Whether remodeling an existing home or building a new one, the rewards from a job well done are manyfold. Whether you do it yourself or hire a professional, you will need to be sure of what you want and how to get it done properly with the least expense.

This book can be used as a text by students in vocational courses, technical colleges, apprenticeship programs, and construction classes in industrial technology programs. The home do it yourselfer will find answers to many questions that pop up in the course of getting a job done, whether over a weekend or over a year's time.

In order to prepare this text, the authors examined courses of study in schools located all over the country. An effort was made to take into consideration the geographic differences and special environmental factors relevant to a particular area. For instance, some locations are not suited for the basement type of house. The slab method of construction for a house foundation is shown as an alternative with its many ramifications and variations. Whether you choose a basement or slab type of house foundation it is important to make sure the proper procedures and materials are used under the right circumstances in order to produce acceptable living quarters once the remodeling is done.

The various chapters of this book include tips, techniques, and step-by-step instructions on how to paint, wallpaper, lay tile, install ceilings, build fences and decks, plan for additions, and more. This comprehensive book on remodeling also includes consumer information on lead-based paints and mold.

No book can be completed without the aid of many people. The Acknowledgments that follow mention some of those who contributed to making this text the most current in design and technology techniques available to the carpenter. We trust you will enjoy using the book as much as we enjoyed writing it.

MARK R. MILLER
REX MILLER
GLENN E. BAKER

Acknowledgments

The authors would like to thank the following manufacturers for their generosity. They furnished photographs, drawings, and technical assistance. Without the donations of time and effort on the part of many people, this book would not have been possible. We hope this acknowledgment of some of the contributors will let you know that the field you are working in or about to enter is one of the best.

Abiti Corporation
American Olean Tile Co.
American Plywood Association
American Standard, Inc.
Armstrong Cork Co.
Bilco Co.
Bird and Son, Inc.
Black & Decker Manufacturing Co.
Closet Maid by Clairson
Dewalt Div., Am. Mach. & Fdry
Duo-fast Corporation
Dutch Boy Paints
Forest Products Laboratory
Formica Corporation
General Products

Grossman Lumber
Jacuzzi, Inc.
Jenn-Air
Kenny Manufacturing Co.
Kohler Company
Lowe's
Millers Falls Div., Ingersoll-Rand Co.
National Lock Hardware Co.
National Oak Flooring Manufacturers Association
NuTone, Inc.
Pattern Magic
Permograin Products
Rockwell International, Power Tools Div.
Sears, Roebuck & Co.
Southern Forest Products Association
Stabila, Inc.
Stanley Tools Co.
Teco Products & Testing Corporation
Town of Amherst, NY
U.S. Gypsum Co.
United Steel Products, Inc.
Valu, Inc.
Weiser Lock, Div. of Norris Industries
Western Wood Products Association

Miller's Guide to Home Remodeling

1

Introduction to Maintenance and Remodeling

IMPORTANCE OF MAINTENANCE

A home is probably the largest investment that you will make in your life. Along with that investment come the upkeep and maintenance problems that are always associated with a home. Let's face it, nothing in this world lasts forever, and lifetime guarantees on products are betting that you will sell the product or lose the receipt by the time the product finally gives out.

A home is made up of a variety of materials and different systems. Most of these materials and systems will probably wear out sooner or later if they are not properly maintained. Proper maintenance can extend the life of any material or system almost indefinitely and will save you a lot of time and money in the future. As the saying goes, "You can pay me now, or you can pay me later." Moreover, if you pay me later, it will be a lot more.

In most cases, routine maintenance, such as painting, cleaning, and adjusting may cost a few dollars or take up a few minutes of your time now. However, if you wait too long, the problem may cause more serious damage which will require the entire replacement of the item and maybe other materials or objects surrounding it. For instance, a loose door hinge will cause the door to scrape when it closes. If you wait too long such that all the screws come loose, the door may bind and crack the door frame and trim, and may even bend the hinge and break the lock set. The door may get dented and will then have to be repaired and refinished. In another instance, caulking on a vinyl floor by a bathtub may crack, and then water will work its way under the floor, lifting and discoloring it (Fig. 1-1). In addition, the wooden floor underneath the tile may rot, and water could start leaking down into the ceiling underneath it. When all is said and done, you will probably have to replace the vinyl floor, the wooden floor underneath it,

and the sheetrock ceiling of the floor beneath it. All this could have been prevented with 10 minutes of your time and a tube of caulk.

IDENTIFYING NEEDED REPAIRS

The first step to doing any maintenance and remodeling is to make a visual inspection of the entire house. This includes the interior and the exterior. You will probably find quite a few things that need to be repaired, so you will have to prioritize them. Repairs dealing with water, mold, asbestos, or anything that may be life-threatening should be dealt with first. In most cases you would like to repair the things you like to do; however, you should always repair the items that have the potential for causing major problems later.

The next step is to list first the items you can repair yourself and then the items for which you will have to seek outside assistance. Again, repair the major problems first, and then repair the little things when your pocketbook cooperates.

DEVELOPING A ROUTINE MAINTENANCE SCHEDULE

Once you have completed all your repairs in your home, the next step is to make a list of all the items that need frequent maintenance. Changing a furnace filter seems like a routine job; however, most people do not keep track of when they change the filter. If a filter is not changed regularly (at least once a month), then the furnace or air conditioner works twice as hard and often burns up motors and/or relays. Attach a paper to the inside of your air conditioning or furnace closet door, and write down the date you replaced the filter, as illustrated in Fig. 1-2. If you always forget, then you may want to

Fig. 1-1 *Caulk between bathtub and floor.*

Fig. 1-2 *Furnace filter replacement date list.*

Fig. 1-3 *Pleated furnace filter (left) and inexpensive spun filter (right).*

buy the more expensive filters that whistle when they are clogged. Further, purchase the filters that pick up over 99% of the particulate matter. Although these filters are much more expensive, they reduce the dust in your house by over 50%, which reduces the number of times you have to dust and vacuum each year. These filters also reduce allergens and molds in the air considerably, which will make your house more enjoyable for family members with allergies. These filters are referred to as *pleated* versus the *spun type* which you can see through (Fig. 1-3).

Other periodic maintenance activities include caulking around window and door frames, cleaning gutters and downspouts, trimming tree branches away from roofing shingles (Figs. 1-4 and 1-5), lubricating locks with graphite powder, repainting or refinishing cracked and exposed wooden surfaces, cleaning mold from around window panes, vacuuming dust from ceiling fan inlets (Fig. 1-6 and 1-7), adjusting and greasing garage door tracks, and power-washing mold and dirt from exterior surfaces. Again, if routine maintenance of a home is followed, potential problems will be eliminated or fixed before they become major

Fig. 1-4 *Tree branches rubbing against roofing shingles.*

Fig. 1-5 *Tree branch scraping roofing shingles and fascia board.*

Fig. 1-6 *Dusty ceiling fan inlets are gray.*

Fig. 1-7 *Clean ceiling fan inlets are darker in color.*

headaches that could cause thousands of dollars' worth of damage and inconvenience to you and your family. Having to reroof your home or purchase a new garage door opener before its time is wasteful and very time-consuming. Remember, most major home repairs are avoidable through simple routine maintenance.

JUSTIFYING REMODELING

The three main reasons why people remodel their homes are to create more space, replace worn out materials, or update their homes to make them more contemporary. Any of these reasons is valid and should increase the value of your home and make it easier to sell when you have to move someday.

Creating space can be as simple as making cabinets under an overhanging countertop (Fig. 1-8), or it can be a major job such as planning room additions to your home. Your budget and time will warrant these endeavors. The main thing to remember is to try to make the addition look as original as possible. If you review the plans and pictures and the floor plan to the new addition seems awkward and unnatural, then everyone will always know it is an addition and it will end up making your home more difficult to sell in the future. Many home buyers are turned off by homes that have had additions because buyers are afraid problems will arise because the two parts of the house are built on different foundations. If there is just no way of adding space without making it look obvious, then it is time to sell your home and buy a new one.

Anytime doors are broken, trim is dented or cracked, sinks and fixtures are corroded and chipped, paint is faded and stained, windows are rotted or rusted, or any other obvious, worn out material or item is noticeable in your home, it is time to replace it and maybe the items around it. In another words, you hate to put a new $100 plumbing fixture on a cracked

Fig. 1-8 *Cabinets built under overhanging countertop.*

porcelain sink with rust stains. You are better off saving your money and probably replacing the cabinet, countertop, sink, and fixture all at once. Typically, it is much easier to do the entire job once than to take it apart each time and redo it again and again. Further, you can break or scratch objects when you take them apart several times.

The last and most important reason to remodel is to make your home look as good as a new home on the market. In most cases, buyers prefer homes where the landscaping has grown in and there is no construction of new homes underway. However, most buyers do not like homes that look dated. The main reason homes look dated is because of the kitchen and bathrooms. All the remaining rooms of a home just have doors and closets. However, kitchens and bathrooms have cabinets, sinks, showers, bathtubs, fixtures, and appliances that change over the years. In addition, these items wear out quickly because they are close to a source of water. Water can rot wood, stain floors, and cause all kinds of damage. Therefore, an older, more established home with an updated kitchen and bathroom would be an ideal find for a potential home buyer.

DETERMINING WHEN NOT TO REMODEL

The number one thing to remember is to not make your house worth more than any other house in the neighborhood, because that will make it more difficult to sell. It is always better to have more expensive homes surround your home. The property values will be higher, and typically upscale neighbors maintain their homes better. If the costs for remodeling are going to be quite extensive and you know you may never recoup the costs because all the homes in your neighborhood are in a state of decline, then you may want to consider moving into another home that better suits your needs.

The second thing to remember, whenever you are remodeling, is that you may want to sell your home someday, so do not make any radical changes to your home that will not appeal to most people. If the remodeling ideas you have in mind are not the norm, you may want to obtain some input from a few local realtors who have sold homes in your neighborhood. Remember, you are not going to live forever, so try to make it easy for one of your relatives to sell your home when you die. You want to be remembered in a positive light.

2
CHAPTER

Safety, Tools, and Equipment

Safety Measures

Protecting the eyes   It is best to wear safety glasses. Make sure your safety glasses are of tempered glass. They will not shatter and cause eye damage. In some instances you should wear goggles. This prevents splinters and other flying objects from entering the eye from under or around the safety glasses. Ordinary glasses aren't always the best, even if they are tempered glass. Just become aware of the possibilities of eye damage whenever you start a new job or procedure. See Fig. 2-1 for a couple of types of safety glasses.

Fig. 2-2   *Face shield.*

Fig. 2-1   *Safety glasses.*

Safety shoes   Sneakers are used only by roofers. Sneakers, sandals, and dress shoes do not provide enough protection for the carpenter on the job. Only safety shoes should be worn on the job.

Gloves   Some types of carpentry work require the sensitivity of bare fingers. Other types do not require hands or fingers to be exposed. In cold or even cool weather, gloves may be in order. Gloves are often needed to protect your hands from splinters and rough materials. It's only common sense to use gloves when you are handling rough materials.

Probably the best gloves for carpentry work are a lightweight type. A suede finish to the leather improves the gripping ability of the gloves. Cloth gloves tend to catch on rough building materials. They may be preferred, however, if you work with short nails or other small objects.

Body protection   Before you go to work on any job, make sure your entire body is properly protected. The hard hat comes in a couple of styles. Under some conditions the face shield is better protection. See Fig. 2-2.

Is your body covered with heavy work clothing? This is the first question to ask before you go onto the job site. Has as much of your body as practical been covered with clothing? Has your head been properly protected? Are your eyes covered with approved safety glasses or a face shield? Are your shoes sturdy, with safety toes and steel soles to protect against nails? Are gloves available when you need them?

General Safety Rules

Some safety procedures should be followed at all times. This applies to carpentry work especially:

- Pay close attention to what is being done.
- Move carefully when you are walking or climbing.
- (Take a look at Fig. 2-3. This type of made-on-the-job ladder can cause trouble.) Use the leg muscles for lifting.
- Move long objects carefully. The end of a carelessly handled 2 × 4 can damage hundreds of dollars worth of glass doors and windows. Keep the workplace neat and tidy. Figure 2-4A shows a cluttered working area. It would be hard to walk along here without tripping. If a dumpster is used for trash and debris, as in Fig. 2-4B, many accidents can be prevented. Sharpen or replace dull tools.
- Disconnect power tools before you adjust them.
- Keep power tool guards in place.
- Avoid interrupting another person who is using a power tool.
- Remove hazards as soon as they are noticed.

Safety on the Job

A safe working site makes it easier to get the job done. Lost time due to accidents puts a building schedule behind. This can cost many thousands of dollars and

Fig. 2-3 *A made-on-the-job ladder.*

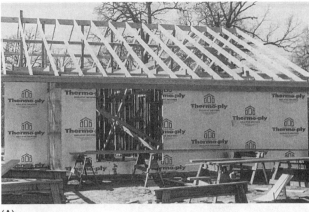

(A)

(B)

Fig. 2-4 *(A) Cluttered work site. (B) A work area can be kept clean if a large dumpster is kept nearby for trash and debris.*

lead to late delivery of the building. If the job is properly organized and safety is taken into consideration, the smooth flow of work is quickly noticed. No one wants to get hurt. Pain is no fun. Safety is just common sense. If you know how to do something safely, it will not take any longer than if you did it in an unsafe manner. Besides, why would you deliberately do something that is dangerous? All safety requires is a few precautions on the job. Safety becomes a habit once you get the proper attitude established in your thinking. Some of these are important habits to acquire:

- Know exactly what is to be done before you start a job.
- Use a tool only when it can be used safely. Wear all safety clothing recommended for the job. Provide a safe place to stand to do the work. Set ladders securely. Provide strong scaffolding.
- Avoid wet, slippery areas.
- Keep the working area as neat as practical.
- Remove or correct safety hazards as soon as they are noticed. Bend protruding nails over. Remove loose boards.

- Remember where other workers are and what they are doing.
- Keep fingers and hands away from cutting edges at all times.
- Stay alert!

Safety Hazards

Carpenters work in unfinished surroundings. While a house is being built, there are many unsafe places around the building site. You have to stand on or climb ladders, which can be unsafe. You may not have a good footing while standing on a ladder. You may not be climbing a ladder in the proper way. Holding onto the rungs of the ladder is very unsafe. You should always hold onto the outside rails of the ladder when climbing.

There are holes that can cause you to trip. They may be located in the front yard where the water or sewage lines come into the building. There may be holes for any number of reasons. These holes can cause you all kinds of problems, especially if you fall into them or turn your ankle.

Fig. 2-5 Even when a house is almost fished, there can still be hazards. Wood left on a roof could slide off and hurt someone, and without the front porch, it is a long step down.

The house in Fig. 2-5 is almost completed. However, if you look closely, you can see that some wood has been left on the garage roof. This wood can slide down and hit a person working below. The front porch has not been poured. This means that stepping out of the front door can be a rather long step. Other debris around the yard can be a source of trouble. Long slivers of flashing can cause trouble if you step on them and they rake your leg. You have to watch your every step around a construction site.

Outdoor work   Much of the time carpentry is performed outdoors. This means you will be exposed to the weather, so dress accordingly. Wet weather increases the accident rate. Mud can make a secure place to stand hard to find. Mud can also cause you to slip if you don't clean it off your shoes. Be very careful when it is muddy and you are climbing on a roof or a ladder.

Tools   Any tool that can cut wood can cut flesh. You have to keep in mind that although tools are an aid to the carpenter, they can also be a source of injury. A chisel can cut your hand as easily as the wood. In fact, it can do a quicker job on your hand than on the wood it was intended for. Saws can cut wood and bones. Be careful with all types of saws, both hand and electric. Hammers can do a beautiful job on your fingers if you miss the nailhead. The pain involved is intensified in cold weather. Broken bones can be easily avoided if you keep your eye on the nail while you're hammering. Besides that, you will get the job done more quickly. And, after all, that's why you are there—to get the job done and do it right the first time. Tools can help you do the job right. They can also cause you injury. The choice is up to you.

To work safely with tools, you should know what they can do and how they do it. The next few pages are designed to help you use tools properly.

USING CARPENTRY TOOLS

A carpenter is lost without tools. This means you have to have some way of containing them. A toolbox is very important. If you have a place to put everything, then you can find the right tool when it is needed. A toolbox should have all the tools mentioned here. In fact, you will probably add more as you become more experienced. Tools have been designed for every task. All it takes is a few minutes with a hardware manufacturer's catalog to find just about everything you'll ever need. If you can't find what you need, the manufacturers are interested in making it.

Measuring Tools

Folding rule   When you are using the folding rule, place it flat on the work. The 0 end of the rule should be exactly even with the end of the space or board to be measured. The correct distance is indicated by the reading on the rule.

A very accurate reading may be obtained by turning the edge of the rule toward the work. In this position, the marked graduations of the face of the rule touch the surface of the board. With a sharp pencil, mark the exact distance desired. Start the mark with the point of the pencil in contact with the mark on the rule. Move the pencil directly away from the rule while you are making the mark.

One problem with the folding rule is that it breaks easily if it is twisted. This happens most commonly when it is being folded or unfolded. The user may not be aware of the twisting action at the time. You should keep the joints oiled lightly. This makes the rule operate more easily.

Pocket tape   Beginners may find the pocket tape (Fig. 2-6) the most useful measuring tool for all types of work. It extends smoothly to full length. It returns quickly to its compact case when the return button is pressed. Steel tapes are available in a variety of lengths. For most carpentry a rule 6, 8, 10, or 12 feet long is used.

Longer tapes are available. They come in 20-, 50-, and 100-foot lengths. See Fig. 2-7. This tape can be extended to 50 feet to measure lot size and the location of a house on a lot. It has many uses around a building site. A crank handle can be used to wind it up once you are finished with it. The hook on the end of the tape makes it easy for one person to use it. Just hook the tape over the end of a board or nail and extend it to your desired length.

Fig. 2-6 *Tape measure.* (Stanley Tools)

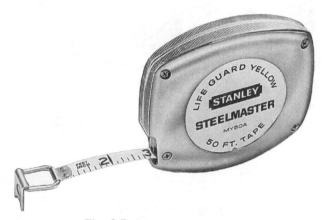

Fig. 2-7 *A longer tape measure.*

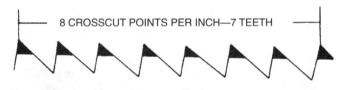

8 CROSSCUT POINTS PER INCH—7 TEETH

6 RIP POINTS PER INCH—5 TEETH

THE NUMBER OF POINTS PER INCH ON A HANDSAW DETERMINES THE FINENESS OR COARSENESS OF CUT. MORE POINTS PRODUCE A FINER CUT.

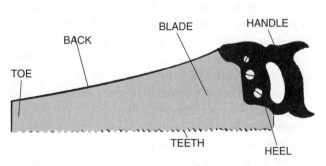

BACK BLADE HANDLE

TOE

TEETH HEEL

Saws

Carpenters use a number of different saws. These saws are designed for specific types of work. Many are misused. They will still do the job, but they would do a better job if used properly. Handsaws take quite a bit of abuse on a construction site. It is best to buy a good-quality saw and keep it lightly oiled.

Standard skew-handsaw This saw has a wooden handle. It has a 22-inch length. A 10-point saw (with 10 teeth per inch) is suggested for crosscutting. Crosscutting means cutting wood *across* the grain. The 26-inch-length, 5½-point saw is suggested for ripping, or cutting *with* the wood grain.

Figure 2-8 shows a carpenter using a handsaw. This saw is used in places where the electric saw cannot be used. Keeping it sharp makes a difference in the quality of the cut and the ease with which it can be used.

Backsaw The backsaw gets its name from the piece of heavy metal that makes up the top edge of the cutting part of the saw. See Fig. 2-9. It has a fine-tooth configuration. This means it can be used to cut cross-grain and leave a smoother finished piece of work. This type of saw is used by finish carpenters who want to cut trim or molding.

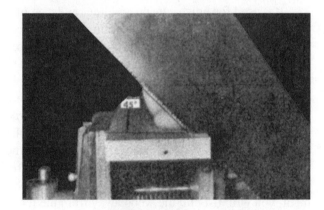

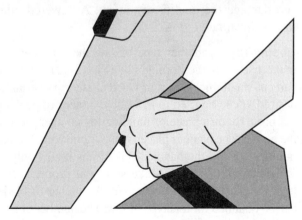

Fig. 2-8 *Using a handsaw.*

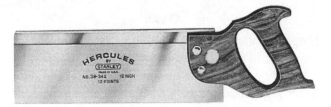

Fig. 2-9 *Backsaw.* (Stanley Tools)

Fig. 2-10 *Miter box.* (Stanley Tools)

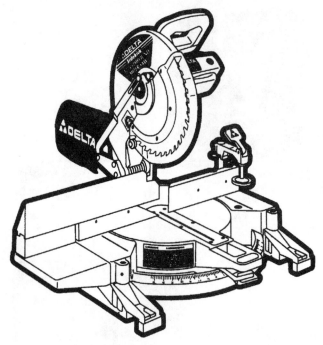

Fig. 2-11 *Powered compound miter saw.* (Delta)

Miter box   As you can see from Fig. 2-10, the miter box has a backsaw mounted in it. This box can be adjusted by using the lever under the saw handle. You can adjust it for the cut you wish. It can cut from 90° to 45°. It is used for finish cuts on moldings and trim materials. The angle of the cut is determined by the location of the saw in reference to the bed of the box. Release the clamp on the bottom of the saw support to adjust the saw to any degree desired. The wood is held with one hand against the fence of the box and the bed. Then the saw is used by the other hand. As you can see from the setup, the cutting should take place when the saw is pushed forward. The backward movement of the saw should be made with the pressure on the saw released slightly. If you try to cut on the backward movement, you will just pull the wood away from the fence and damage the quality of the cut. A powered miter saw is illustrated in Fig. 2-11.

Coping saw   Another type of saw the carpenter can make use of is the coping saw (Fig. 2-12). This one can cut small thicknesses of wood at any curve or angle desired. It can be used to make sure a piece of paneling fits properly or a piece of molding fits another piece in the corner. The blade is placed in the frame with the teeth pointing toward the handle. This means it cuts only on the downward stroke. Make sure you properly support the piece of wood being cut. A number of blades can be obtained for this type of saw. The number of teeth in the blade determines the smoothness of the cut.

Fig. 2-12 *Coping saw.* (Stanley Tools)

Hammers and Other Small Tools

There are a number of different types of hammers. The one the carpenter uses is the *claw* hammer. It has claws that can extract nails from wood if they have been put in the wrong place or have bent while being driven. Hammers can be bought in 20-ounce, 24-ounce, 28-ounce, and 32-ounce weights for carpentry work. Most carpenters prefer a 20-ounce. You have to work with a number of different weights to find out which will work best for you. Keep in mind that the hammer should be of tempered steel. If the end of the hammer has a tendency to splinter or chip off when it hits a nail, the pieces can hit you in the eye or elsewhere, causing serious damage. It is best to wear safety glasses whenever you use a hammer.

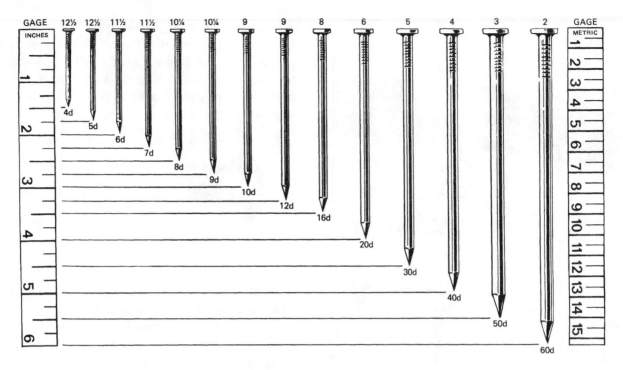

Nails are driven by hammers. Figure 2-13 shows the gauge, inch, and penny relationships for the common box nail. The *d* after the number means *penny*. This is a measuring unit inherited from the English in the Colonial days. There is little or no relationship between penny and inches. If you want to be able to talk about it intelligently, you'll have to learn both inches and penny. The gauge is nothing more than the American Wire Gage number for the wire that the nails were made from originally. Finish nails have the same measuring unit (penny) but do not have the large, flat heads.

Nail set   Finish nails are driven below the surface of the wood by a nail set. The nail set is placed on the head of the nail. The large end of the nail set is struck by the hammer. This causes the nail to go below the surface of the wood. Then the hole left by the countersunk nail is filled with the wood filler and finished off with a smooth coat of varnish or paint. Figure 2-14 shows the nail set and its use.

The carpenter would be lost without a hammer. See Fig. 2-15. Here, the carpenter is placing sheathing on rafters to form a roof base. The hammer is used to

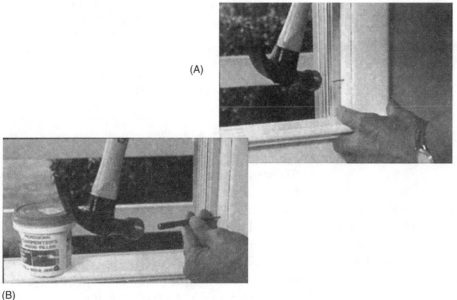

(A)

(B)

Fig. 2-14 (A) Driving a nail with a hammer. (B) Finishing the job with a nail set to make sure the hammer doesn't leave an impression in the soft wood of the window frame.

Fig. 2-15 *Putting on roof sheathing. The carpenter is using a hammer to drive the board into place.*

drive the boards into place, since they have to overlap slightly. Then the nails are driven by the hammer also.

In some cases a hammer will not do the job. The job may require a hatchet. See Fig. 2-16. This device can be used to pry and to drive. It can pry boards loose

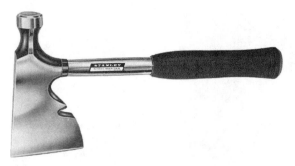

Fig. 2-16 *Hatchet.* (Stanley Tools)

when they are improperly installed. It can sharpen posts to be driven at the site. The hatchet can sharpen the ends of stakes for staking out the site. It can also withdraw nails. This type of tool can be used to drive stubborn sections of a wall into place when they are erected for the first time. The tool has many uses.

Scratch awl An awl is a handy tool for a carpenter. It can be used to mark wood with a scratch mark and to produce pilot holes for screws. Once it is in your toolbox, you can think of a hundred uses for it. Since it does have a very sharp point, it is best to treat it with respect. See Fig. 2-17.

Fig. 2-17 *Scratch awl.* (Stanley Tools)

Wrecking bar This device (Fig. 2-18) has a couple of names, depending on which part of the country you are in at the time. It is called a wrecking bar in some parts and a crowbar in others. One end has a chisel-sharp flat surface to get under boards and pry them loose. The other end is hooked so that the slot in the end can pull nails with the leverage of the long handle. This specially treated steel bar can be very helpful in prying away old and unwanted boards. It can be used to help give leverage when you are putting a wall in place and making it plumb. This tool has many uses for the carpenter with ingenuity.

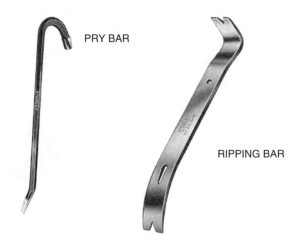

PRY BAR

RIPPING BAR

Fig. 2-18 *Wrecking bars.* (Stanley Tools)

Screwdrivers The screwdriver is an important tool for the carpenter. It can be used for many things other than turning screws. There are two types of screwdrivers. The standard type has a straight slot-fitting blade at its end. This type is the most common of screwdrivers. The Phillips head screwdriver has a cross or X on the end to fit a screw head of the same design. Figure 2-19 shows the two types of screwdrivers.

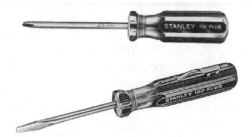

Fig. 2-19 *Two types of screwdrivers.*

Squares

To make corners meet and standard sizes of materials fit properly, you must have things square. That calls for a number of squares to check that the two walls or two pieces come together at a perpendicular.

Try square The try square can be used to mark small pieces for cutting. If one edge is straight and the handle part of the square (Fig. 2-20) is placed against this straightedge, then the blade can be used to mark the wood perpendicular to the edge. This comes in handy when you are cutting 2 × 4s and want them to be square.

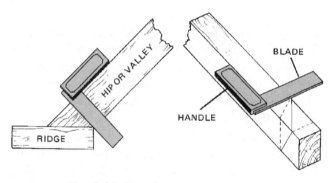

Fig. 2-20 *Use of a try square.* (Stanley Tools)

Framing square The framing square is a very important tool for the carpenter. It allows you to make square cuts in dimensional lumber. This tool can be used to

lay out rafters and roof framing. See Fig. 2-21. It is also used to lay out stair steps.

Later in this book you will see a step-by-step procedure for using the framing square. The tools are described as they are called for in actual use.

Bevel A bevel can be adjusted to any angle to make cuts at the same number of degrees. See Fig. 2-22. Note how the blade can be adjusted. Now take a look at Fig. 2-23. Here you can see the overhang of rafters.

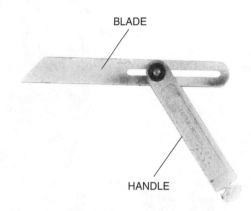

Fig. 2-22 *Bevel.* (Stanley Tools)

If you want the ends to be parallel with the side of the house, you can use the bevel to mark them before they are cut off. Simply adjust the bevel so the handle is on top of the rafter and the blade fits against the soleplate below. Tighten the screw and move the bevel down the rafter to where you want the cut. Mark the angle along the blade of the bevel. Cut along the mark, and you have what you see in Fig. 2-23. It is a good device for transferring angles from one place to another.

Chisel Occasionally you may need a wood chisel. It is sharpened on one end. When the other end is struck with a hammer, the cutting end will do its job. That is, of course, if you have kept it sharpened. See Fig. 2-24.

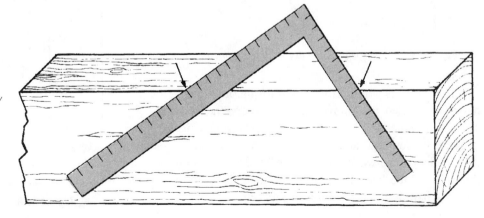

Fig. 2-21 *Framing square.* (Stanley Tools)

Fig. 2-23 *Rafter overhang cut to a given angle.*

metal tip on the handle so the force of the hammer blows will not chip the handle. Other applications are up to you, the carpenter. You'll find many uses for the chisel in making things fit.

Plane Planes (Fig. 2-25) are designed to remove small shavings of wood along a surface. One hand holds the knob in front, and the other hand holds the handle in back. The blade is adjusted so that only a small sliver of wood is removed each time the plane is passed over the wood. It can be used to make sure that doors and windows fit properly. It can be used for any number of wood smoothing operations.

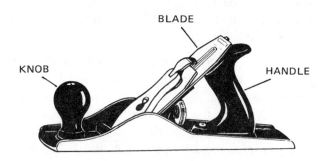

Fig. 2-25 *Smooth plane.* (Stanley Tools)

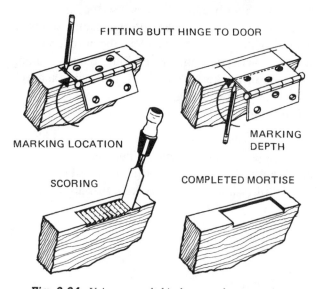

Fig. 2-24 *Using a wood chisel to complete a mortise.*

The chisel is commonly used in fitting or hanging doors. It is used to remove the area where the door hinge fits. Note how it is used to score the area (Fig. 2-24); it is then used at an angle to remove the ridges. A great deal of the work with the chisel is done by using the palm of the hand as the force behind the cutting edge. A hammer can be used. In fact, chisels have a

Dividers and compass Occasionally a carpenter must draw a circle. This is done with a compass. The compass shown in Fig. 2-26A can be converted to a divider by removing the pencil and inserting a straight steel pin. The compass has a sharp point that fits into the wood surface. The pencil part is used to mark the circle circumference. It is adjustable to various radii.

The dividers in Fig. 2-26A have two points made of hardened metal. They are adjustable. It is possible

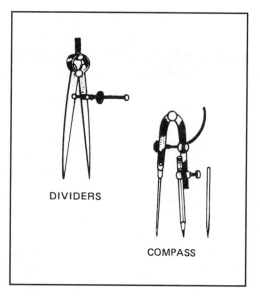

Fig. 2-26A *Dividers and compass.*

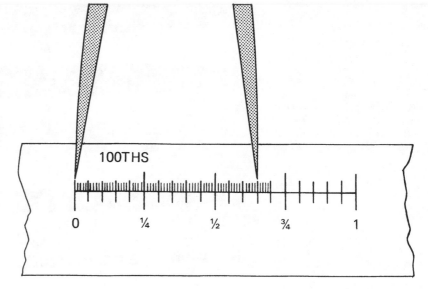

Fig. 2-26B *Dividers being used to transfer hundredths of an inch.*

100THS

0   ¼   ½   ¾   1

to use them to transfer a given measurement from the framing square or measuring device to another location. See Fig. 2-26B.

Level   To have things look as they should, a level is necessary. There are a number of sizes and shapes available. The one shown in Fig. 2-27B is the most common type used by carpenters. The bubbles in the glass tubes tell you if the level is obtained. In Fig. 2-27A the carpenter is using the level to make sure the window is in properly, before nailing it into place permanently.

If the vertical and horizontal bubbles are lined up between the lines, then the window is plumb, or vertical.

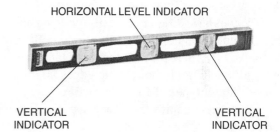

HORIZONTAL LEVEL INDICATOR

VERTICAL INDICATOR   VERTICAL INDICATOR

Fig. 2-27B   *A commonly used type of level.* (Stanley Tools)

A plumb bob is a small, pointed weight. It is attached to a string and dropped from a height. If the bob is just above the ground, it will indicate the vertical direction by its string. Keeping windows, doors, and frames square and level makes a difference in fitting. It is much easier to fit prehung doors in a frame that is square. When it comes to placing panels of 4-foot × 8-foot plywood sheathing on a roof or on walls, squareness can make a difference as to fit. Besides, a square fit and a plumb door and window look better than those that are a little off. Figure 2-27C shows three plumb bobs.

Fig. 2-27A   *Using a level to make sure a window is placed properly before nailing.* (Andersen)

PLUMB BOBS

Fig. 2-27C   *Plumb bobs.* (Stanley Tools)

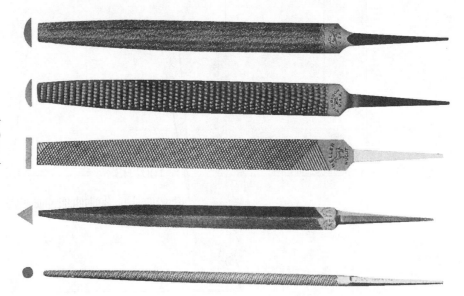

Fig. 2-28 *Wood and cabinet files: (A) Half-round; (B) rasp; (C) flat; (D) triangular; and (E) round.* (Millers Falls Division, a division of Ingersol-Rand Co.)

Files A carpenter finds use for a number of types of files. The files have different surfaces for doing different jobs. Tapping out a hole to get something to fit may be just the job for a file. Some files are used for sharpening saws and touching up tool cutting edges. Figure 2-28 shows different types of files. Other files may also be useful. You can acquire them later as you develop a need for them.

Clamps C clamps are used for many holding jobs. They come in handy when you are placing kitchen cabinets by holding them in place until screws can be inserted and properly seated. This type of clamp can be used for an extra hand every now and then, when two hands aren't enough to hold a combination of pieces until you can nail them. See Fig. 2-29.

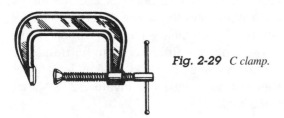

Fig. 2-29 *C clamp.*

Cold chisel It is always good to have a cold chisel around. It is very much needed when you can't remove a nail. Its head may have broken off, and the nail must be removed. The chisel can cut the nail and permit the separation of the wood pieces. See Fig. 2-30.

Fig. 2-30 *Cold chisel.* (Stanley Tools)

If a chisel of this type starts to "mushroom" at the head, you should remove the splintered ends with a grinder. Hammering on the end can produce a mushrooming effect. These pieces should be taken off since they can easily fly off when hit with a hammer. That is another reason for using eye protection when you are using tools.

Caulking gun In times of energy crisis, the caulking gun gets plenty of use. It is used to fill in around windows and doors and everywhere there may be an air leak. There are many types of caulk being made today.

This gun is easily operated. Insert the cartridge and cut its tip to the shape you want. Puncture the thin plastic film inside. A bit of pressure will cause the caulk to come out the end. The long rod protruding from the end of the gun is turned over. This is so the serrated edge will engage the hand trigger. Remove the pressure from the cartridge when you are finished. Do this by rotating the rod so that the serrations are not engaged by the trigger of the gun.

Power Tools

The carpenter uses many power tools to aid in getting the job done. The quicker the job is done, the more valuable the work of the carpenter becomes. This is called productivity. The more you are able to produce, the more valuable you are. This means the contractor can make money on the job. This means you can have a job the next time there is a need for a good carpenter. Power tools make your work go faster. They also help you to do a job without getting fatigued. Many tools have been designed with you in mind. They are portable and operate from an extension cord.

The extension cord should be the proper size to take the current needed for the tool being used. See Table 2-1. Note how the distance between the outlet and the tool using the power is critical. If the distance is great, then the wire must be larger in size to handle the current without too much loss. The higher the number of the wire, the smaller the diameter of the wire. The larger the size of the wire (diameter), the more current it can handle without dropping the voltage.

Some carpenters run an extension cord from the house next door for power before the building site is furnished with power. If the cord is too long or has the wrong size wire, it drops the voltage below 115. This means the saws or other tools using electricity will draw more current and therefore drop the voltage more. Every time the voltage is dropped, the device tries to obtain more current. This becomes a self-defeating phenomenon. You wind up with a saw that has little cutting power. You may have a drill that won't drill into a piece of wood without stalling. Of course the damage done to the electric motor in some cases is irreparable. You may have to buy a new saw or drill. Double-check Table 2-1 for the proper wire size in your extension cord.

Portable saw   This is the most often used and abused of the carpenter's equipment. The electric portable saw, such as the one shown in Fig. 2-31, is used to cut all 2 × 4s and other dimensional lumber. It is used to cut off rafters. This saw is used to cut sheathing for roofs. It is used for almost every sawing job required in carpentry.

This saw has a guard over the blade. The guard should always be left intact. Do not remove the saw guard. If not held properly against the wood being cut, the saw can kick back and into your leg.

You should always wear safety glasses when using this saw. The sawdust is thrown in a number of directions,

Fig. 2-31 *Portable power saw—the favorite power tool of every carpenter. Note the blade should not extend more than $1/8$ inch below the wood being cut. Also note the direction of the blade rotation.*

and one of these is straight up toward your eyes. If you are watching a line where you are cutting, you definitely should have on glasses.

Table saw   If the house has been enclosed, it is possible to bring in a table saw to handle the larger cutting jobs. See Fig. 2-32. You can do ripping a little more safely with this type of saw because it has a rip fence. If a push stick is used to push the wood through and past the blade, it is safe to operate. Do not remove the safety guard. This saw can be used for both crosscut and rip. The blade is lowered or raised to the thickness of the wood. It should protrude about ¼ to ½ inch above the wood being cut. This saw usually requires a 1-horsepower motor. This means it will draw about 6.5 amperes to run and over 35 amperes to start. It is best not to run the saw on an extension cord. It should be wired directly to the power source with circuit breakers installed in the line.

Table 2-1 *Size of Extension Cords for Portable Tools*

Cord Length, Feet	Full-Load Ratings of the Tool (in Amperes) at 115 Volts					
	0 to 2.0	2.10 to 3.4	3.5 to 5.0	5.1 to 7.0	7.1 to 12.0	12.1 to 16.0
	Wire Size (AWG)					
25	18	18	18	16	14	14
50	18	18	18	16	14	12
75	18	18	16	14	12	10
100	18	16	14	12	10	8
200	16	14	12	10	8	6
300	14	12	10	8	6	4
400	12	10	8	6	4	4
500	12	10	8	6	4	2
600	10	8	6	4	2	2
800	10	8	6	4	2	1
1000	8	6	4	2	1	0

If the voltage is lower than 115 volts at the outlet, have the voltage increased or use a much larger cable than listed.

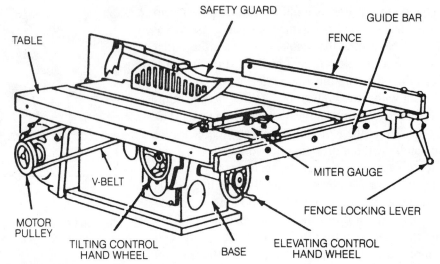

TABLE

SAFETY GUARD

GUIDE BAR

FENCE

MITER GAUGE

FENCE LOCKING LEVER

ELEVATING CONTROL HAND WHEEL

BASE

TILTING CONTROL HAND WHEEL

MOTOR PULLEY

V-BELT

Fig. 2-32 *Table saw.* (Power Tool Division, Rockwell International)

Radial arm saw This type of saw is brought in only if the house can be locked up at night. The saw is expensive and too heavy to be moved every day. It should have its own circuit. The saw will draw a lot of current when it hits a knot while cutting wood. See Fig. 2-33.

Fig. 2-33 *Radial arm saw.* (DeWalt)

In this model the moving saw blade is pulled toward the operator. In the process of being pulled toward you, the blade rotates so that it forces the wood being cut against the bench stop. Just make sure your left hand is in the proper place when you pull the blade back with your right hand. It takes a lot of care to operate a saw of this type. The saw works well for cutting large-dimensional lumber. It will crosscut or rip. This saw will also do miter cuts at almost any angle. Once you become familiar with it, the saw can be used to bevel crosscut, bevel miter, bevel rip, and even cut circles. However, it does take practice to develop some degree of skill with this saw.

Router The router has a high-speed type of motor. It will slow down when overloaded. It takes the beginner some time to adjust to feeding the router properly. If you feed it too fast, it will stall or burn the edge you're routing. If you feed it too slowly, it may not cut the way you wish. You will have to practice with this tool for some time before you will be ready to use it to make furniture. It can be used for routing holes where needed. It can be used to take the edges off laminated plastic on countertops. Use the correct bit, though. This type of tool can be used to the extent of the carpenter's imagination. See Fig. 2-34.

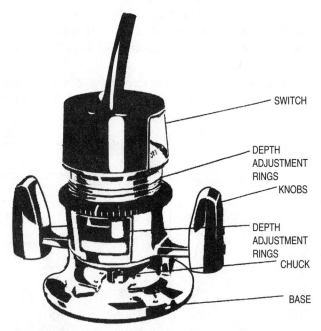

SWITCH

DEPTH ADJUSTMENT RINGS

KNOBS

DEPTH ADJUSTMENT RINGS

CHUCK

BASE

Fig. 2-34 *The handheld router has many uses in carpentry.*

Saw blades There are a number of saw blades available for the portable, table, or radial saw. They may be

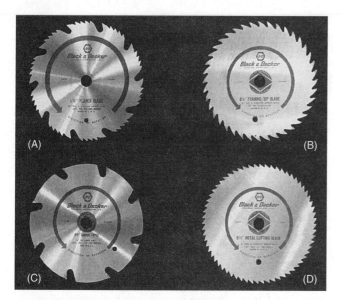

Fig. 2-35 *Saw blades. (A) Planer blade; (B) framing rip blade; (C) carbide-tipped; (D) metal cutting blade.* (Black & Decker)

Fig. 2-36 *Saber saw.*

standard steel types, or they may be carbide-tipped. Carbide-tipped blades tend to last longer. See Fig. 2-35.

Combination blades (those that can be used for both crosscut and rip) with a carbide tip give a smooth finish. They come in 7- to 7¼-inch diameter with 24 teeth. The arbor hole for mounting the blade on the saw is ¾ to ⅝ inch. A safety combination blade is also made in 10-inch-diameter size with 10 teeth and the same arbor hole sizes as the combination carbide-tipped blade.

The planer blade is used to crosscut, rip, or miter hard- or softwoods. It is 6½ or 10 inches in diameter with 50 teeth. It too can fit anything from ¾- to ⅝-inch arbors.

If you want a smooth cut on plywood without the splinters that plywood can generate, you had better use a carbide-tipped plywood blade. It is equipped with 60 teeth and can be used to cut plywood, Formica, or laminated countertop plastic. It can also be used for straight cutoff work in hard- or softwoods. Note the shape of the saw teeth to get some idea as to how each is designed for a specific job. You can identify these after using them for some time. Until you can, mark them with a grease pencil or marking pen when you take them off. A Teflon-coated blade works better when you are cutting treated lumber.

Saber saw The saber saw has a blade that can be used to cut circles in wood. See Fig. 2-36. It can be used to cut around any circle or curve. If you are making an inside cut, it is best to drill a starter hole first. Then insert the blade into the hole and follow your mark. The saber saw is especially useful in cutting out holes for heat ducts in flooring. Another use for this type of saw is to cut holes in roof sheathing for pipes and other protrusions. The saw blade is mounted so that it cuts on the upward stroke. With a fence attached, the saw can also do ripping.

Drill The portable power drill is used by carpenters for many tasks. Holes must be drilled in soleplates for anchor bolts. Using an electric power drill (Fig. 2-37A) is faster and easier than drilling by hand. This drill is capable of drilling almost any size hole through dimensional lumber. A drill bit with a carbide tip enables the carpenter to drill in concrete as well as bricks. Carpenters use this type of masonry hole to insert anchor bolts in concrete that has already hardened. Electrical boxes have to be mounted in drilled holes in brick and concrete. The job can be made easier and can be more efficiently accomplished with the portable power hand drill.

Fig. 2-37A *Handheld portable drill.*

The drill has a tough, durable plastic case. Plastic cases are safer when used where there is electrical work in progress.

Carpenters are now using cordless electric drills (Fig. 2-37B). Cordless drills can be moved about the job without the need for extension cords. Improved battery technology has made the cordless drills almost as powerful as regular electric drills. The cordless drill has numbers on the chuck to show the power applied to the shaft. Keep in mind that the higher the number, the greater the torque. At low power settings, the chuck will slip when the set level of power is reached. This allows the user to set the drill to drive screws.

Fig. 2-37B *A cordless hand drill with variable torque.*

Figure 2-37C shows a cordless drill and a cordless saw. This cordless technology is now used by carpenters and do-it-yourselfers. Cordless tools can be obtained in sets that use the same charger system (Fig. 2-37D). An extra set of batteries should be kept charging at all

Fig. 2-37C *A cordless drill and a cordless saw using matching batteries.*

Fig. 2-37D *One charger can be used to charge saw and drill batteries of the same voltage.*

times and then swapped out for the discharged ones. This way no time is lost waiting for the battery to reach full charge. Batteries for cordless tools are rated by battery voltage. High voltage gives more power than low voltage.

As a rule, battery-powered tools do not give the full power of regular tools. However, most jobs don't require full power. Uses for electric drills are limited only by the imagination of the user. The cordless feature is very handy when you are mounting countertops on cabinets. Sanding disks can be placed in the tool and used for finishing wood. Wall and roof parts are often screwed in place rather than nailed. Using the drill with special screwdriver bits can make the job faster than nailing.

Sanders The belt sander shown in Fig. 2-38 and the orbital sanders shown in Fig. 2-39A and B can do almost any required sanding job. The carpenter needs the sander occasionally. It helps align parts properly, especially those that don't fit by just a small amount. The sander can be used to finish off windows, doors, counters, cabinets, and floors. A larger model of the belt sander is used to

Fig. 2-38 *Belt sander.* (Black & Decker)

(A)

(B)

Fig. 2-39 *Orbital sanders: (A) dual-action and (B) single-action.*
(Black & Decker)

sand floors before they are sealed and varnished. The orbital or vibrating sanders are used primarily to put a very fine finish on a piece of wood. Sandpaper is attached to the bottom of the sander. The sander is held by hand over the area to be sanded. The operator has to remove the sanding dust occasionally to see how well the job is progressing.

Nailers One of the greatest tools the carpenter has acquired recently is the nailer. See Fig. 2-40. It can drive nails or staples into wood better than a hammer. The nailer is operated by compressed air. The staples and nails are especially designed to be driven by the machine. See Tables 2-2 and 2-3 for the variety of fasteners used with this type of machine. The stapler or nailer can also be used to install siding or trim around a window.

The tool's low air pressure requirements (60 to 90 pounds per square inch) allow it to be moved from

Fig. 2-40 *Air-powered nailer.* (Duo-Fast)

place to place. Nails for this machine are from 6*d* to 16*d*. It is magazine-fed for rapid use. Just pull the trigger.

FOLLOWING CORRECT SEQUENCES

One of the important things a carpenter must do is to follow a sequence. Once you start a job, the sequence has to be followed properly to arrive at a completed house in the minimum amount of time.

Preparing the Site

Preparing the site may be expensive. There must be a road or street. In most cases the local ordinances require a sewer. In most locations the storm sewer and the sanitary sewer must be in place before building starts. If a sanitary sewer is not available, you should plan for a septic tank for sewage disposal.

Figure 2-41 shows a sewer project in progress. This shows a street being extended. The storm sewer lines are visible, as is the digger. Trees had to be removed first by a bulldozer. Once the sewer lines are in, the roadbed or street must be properly prepared. Figure 2-42 shows the building of a street. Proper drainage is very important. Once the street is in and the curbs are poured, it is time to locate the house.

Table 2-2 *Fine Wire Staples for a Pneumatic Staple Driver*

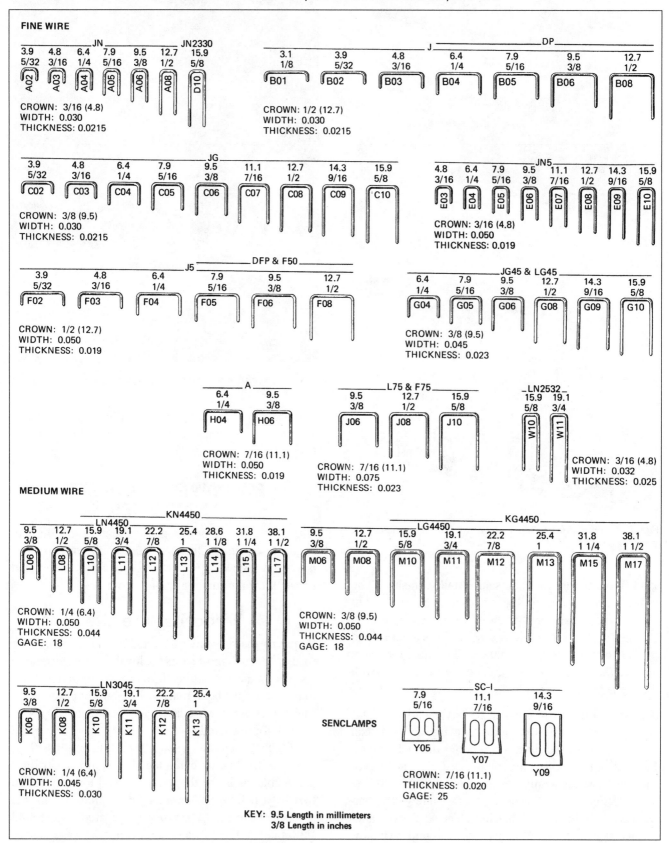

Table 2-3 Seven-Digit Nail Ordering System

1st Digit: Diameter, Inches	2d Digit: Head	3d and 4th Digits: Length, Inches	5th Digit: Point	6th Digit: Wire Chem. and Finish	7th Digit: Finish
A 0.0475	**A** Brad	**08** 1/2	**A** Diam.-reg.	**A** Std. carbon-galv.	**A** Plain
D 0.072	**C** Flat	**11** 3/4	**E** Chisel	**E** Std. carb. "Weatherex" galv.	**B** Sencote
E 0.0915	**E** Flat/ring shank	**13** 1			**C** Painted
G 0.113		**15** 1 1/4		**G** Stainless steel std. tensile	**D** Painted and sencote
H 0.120	**F** Flat/screw shank	**17** 1 1/2			
J 0.105		**19** 1 3/4		**H** Hardened high-carbon bright basic	
K 0.131	**Y** Slight-headed pin	**20** 1 7/8			
U 0.080		**21** 2		**P** Std. carbon bright basic	
	Z Headless pin	**22** 2 1/8			
		23 2 1/4			
		24 2 3/8			
		25 2 1/2			
		26 2 3/4			
		27 3		EXAMPLE: 10 1/4 ga. (K), flat head (C),	
		28 3 1/4		KC25AAA—2 1/2" (25), regular point (A), std. carb.	
		29 3 1/2		galvanized (A), plain, or uncoated (A) Senco-Nail	

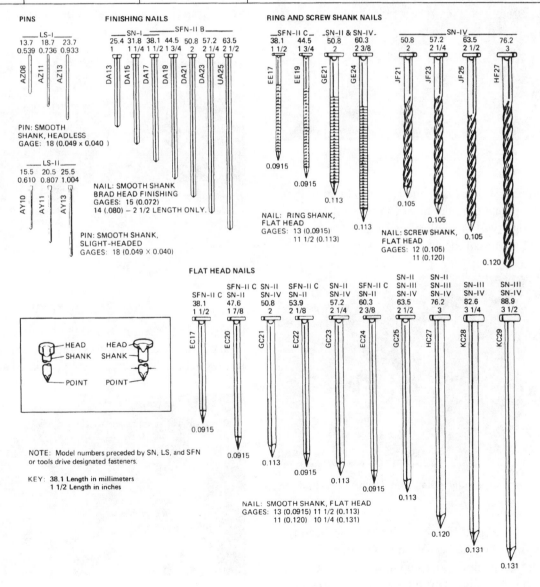

PINS

LS-I
13.7 18.7 23.7
0.539 0.736 0.933
AZ08 AZ11 AZ13

PIN: SMOOTH SHANK, HEADLESS
GAGE: 18 (0.049 x 0.040)

LS-II
15.5 20.5 25.5
0.610 0.807 1.004
AY10 AY11 AY13

PIN: SMOOTH SHANK, SLIGHT-HEADED
GAGES: 18 (0.049 × 0.040)

FINISHING NAILS

SN-I _____ SFN-II B
25.4 31.8 38.1 44.5 50.8 57.2 63.5
1 1 1/4 1 1/2 1 3/4 2 2 1/4 2 1/2
DA13 DA15 DA17 DA19 DA21 DA23 UA25

NAIL: SMOOTH SHANK BRAD HEAD FINISHING
GAGES: 15 (0.072)
14 (.080) — 2 1/2 LENGTH ONLY.

RING AND SCREW SHANK NAILS

SFN-II C
38.1 44.5
1 1/2 1 3/4
EE17 EE19
0.0915 0.0915

SN-II & SN-IV
50.8 60.3
2 2 3/8
GE21 GE24
0.113 0.113

NAIL: RING SHANK, FLAT HEAD
GAGES: 13 (0.0915)
11 1/2 (0.113)

SN-IV
50.8 57.2 63.5 76.2
2 2 1/4 2 1/2 3
JF21 JF23 JF25 HF27
0.105 0.105 0.105 0.120

NAIL: SCREW SHANK, FLAT HEAD
GAGES: 12 (0.105)
11 (0.120)

FLAT HEAD NAILS

SFN-II C
38.1
1 1/2
EC17
0.0915

SFN-II C SN-II
47.6
1 7/8
EC20
0.0915

SN-II SN-IV
50.8
2
GC21
0.113

SFN-II C SN-II
53.9
2 1/8
EC22
0.0915

SN-II SN-IV
57.2
2 1/4
GC23
0.113

SFN-II C SN-II
60.3
2 3/8
EC24
0.0915

SN-II SN-III SN-IV
63.5
2 1/2
GC25
0.113

SN-II SN-III SN-IV
76.2
3
HC27
0.120

SN-III SN-IV
82.6
3 1/4
KC28
0.131

SN-III SN-IV
88.9
3 1/2
KC29
0.131

NAIL: SMOOTH SHANK, FLAT HEAD
GAGES: 13 (0.0915) 11 1/2 (0.113)
11 (0.120) 10 1/4 (0.131)

HEAD — HEAD
SHANK — SHANK
POINT — POINT

NOTE: Model numbers preceded by SN, LS, and SFN or tools drive designated fasteners.

KEY: **38.1 Length in millimeters**
1 1/2 Length in inches

Fig. 2-41 *Street being extended for a new subdivision.*

Fig. 2-42 *The beginning of a street.*

Figure 2-43 shows how the curb has been broken and the telephone terminal box installed in the weeds. Note the stake with a small piece of cloth on it. This marks the location of the site.

Fig. 2-43 *Locating a building site and removing the curb for the driveway.*

As you can see in Fig. 2-44, the curb has been removed. A gravel bed has been put down for the driveway.

The sewer manhole sticks up in the driveway. The basement has been dug. Dirt piles around it show how deep the basement really is. However, a closer look

Fig. 2-44 *Dirt from the basement excavation is piled high around a building site.*

shows that the hole isn't too deep. That means the dirt will be pushed back against the basement wall to form a higher level for the house. This will provide drainage away from the house when finished. See Fig. 2-45 for a look at the basement hole.

Fig. 2-45 *Hole for a basement.*

The Basement

In Fig. 2-46 the columns and the foundation wall have been put up. The basement is prepared in this case with courses of block with brick on the outside. This basement

Fig. 2-46 *The columns and foundation walls will help support the floor parts.*

appears to be more of a crawl space under the first floor than a full stand-up basement.

Once the basement is finished and the floor joists have been placed, the flooring is next.

The Floor

Once the basement or foundation has been laid for the building, the next step is to place the floor over the joists. Note in Fig. 2-47 that the grooved flooring is laid in large sheets. This makes the job go faster and reinforces the floor.

Fig. 2-47 Carpenters are laying plywood subflooring with tongue-and-groove joints. This is stronger. (American Plywood Association)

Wall Frames

Once the floor is in place and the basement entrance hole has been cut, the floor can be used to support the wall frame. The 2 × 4s or 2 × 6s for the framing can be placed on the flooring and nailed. Once together, they are pushed into the upright position, as in Fig. 2-48.

Fig. 2-48 Wall frames are erected after the floor frame is built.

For a two-story house, the second floor is placed on the first-story wall supports. Then the second-floor walls are nailed together and raised into position.

Sheathing

Once the sheathing is on and the walls are upright, it is time to concentrate on the roof. See Fig. 2-49. The rafters are cut and placed into position and nailed firmly. See Fig. 2-50. They are reinforced by the proper horizontal bracing. This makes sure they are properly designed for any snow load or other loads that they may experience.

Fig. 2-49 Beginning construction of the roof structure. (Georgia-Pacific)

Fig. 2-50 Framing and supports for rafters.

Roofing

The roofing is applied after the siding is on and the rafters are erected. The roofing is completed by applying the proper underlayment and then the shingles. If asphalt shingles are used, the procedure is

Fig. 2-51 *Fiberboard sheathing over the wall frame.*

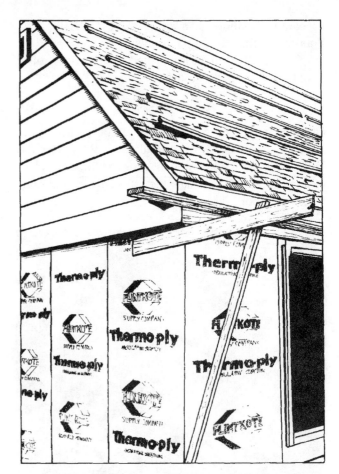

Fig. 2-52 *Siding applied on the top left side of the building.*

Fig. 2-53 *Siding applied to building. Note the pattern of the staples.*

slightly different from that for wooden shingles. Figure 2-51 shows the sheathing in place and ready for the roofing.

Siding

After the roofing, the finishing job will have to be undertaken. The windows and doors are in place. Finish touches are next. The plumbing and drywall may already be in. Then the siding has to be installed. In some cases, of course, it may be brick. This calls for bricklayers to finish up the exterior. Otherwise the carpenter places siding over the walls. Figure 2-52 shows the beginning of the siding at the top left of the picture.

Figure 2-53 shows how the siding has been held in place with a stapler. The indentations in the wood show a definite pattern. The siding is nailed to the nail base underneath after a coating of tar paper (felt paper in some parts of the country) is applied to the nail base or sheathing.

Finishing

Exterior finishing requires a bit of caulking with a caulking gun. Caulk is applied to the siding that butts the windows and doors.

Finishing the interior can be done at a more leisurely pace once the exterior is enclosed. The plumbing and electrical work have to be done before the drywall or plaster is applied. Once the wallboard has been finished, the trim can be placed around the edges of the walls, floors, windows, and doors. The flooring can be applied after the finishing of the walls and ceiling. The kitchen cabinets must be installed before the kitchen flooring. There is a definite sequence to all these operations.

As you can imagine, it would be impossible to place roofing on a roof that wasn't there. It takes plan-

ning and following a sequence to make sure the roof is there when the roofing crew comes around to nail the shingles in place. The water must be there before you can flush the toilets. The electricity must be hooked up before you can turn on a light. These are reasonable things. All you have to do is sit down and plan the whole operation before starting. Planning is the key to sequencing. Sequencing makes it possible for everyone to be able to do a job at the time assigned to it.

The Laser Level

The need for plumb walls and level moldings as well as various other points straight and level is paramount in house building. It is difficult in some locations to establish a reference point to check for level windows, doors, and roofs, as well as ceilings and steps.

The laser level (Fig. 2-54) has eliminated much of this trouble in house building. This simple, easy-to-use tool is accurate to within 1/8 inch in 150 feet, and it has become less expensive recently so that even do-it-yourselfers can rent or buy one.

Fig. 2-54 *Laserspirit level moves 360° horizontally and 360° vertically with the optional lens attachment. It sets up quickly and simply with only two knobs to adjust.* (Stabila®)

The laser level can generate a vertical reference plane for positioning a wall partition or for setting up forms. See Fig. 2-55. It can produce accurate height gauging and alignment of ceilings, moldings, and horizontal planes and can accurately locate doorways,

Fig. 2-55 *The laser level can be used to align ceilings, moldings, and horizontal planes; it produces accurate locations for doorways, windows, and thresholds for precision framing and finishing.* (Stabila®)

windows, and thresholds for precision framing and finishing. See Fig. 2-56. The laser level can aid in leveling floors, both indoors and outdoors. It can be used to check stairs, slopes, and drains. The laser beam is easy to use and accurate in locating markings for roof pitches, and it works well in hard-to-reach situations. See Fig. 2-57. The laser bean is generated by two AAA alkaline batteries that will operate for up to 16 hours.

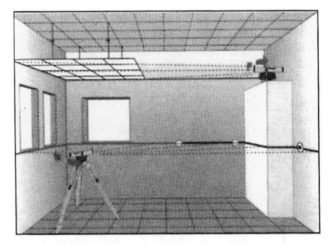

Fig. 2-56 *The laser level can be used for indoor or outdoor leveling of floors, stairs, slopes, drains and ceilings, and moldings around the room.* (Stabila®)

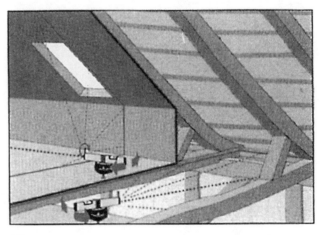

Fig. 2-57 *The laser beam is used to provide easy and accurate location markings on pitches and in hard-to-reach situations.* (Stabila®)

The combination laser and spirit level quickly and accurately lays out squares and measures plumb. No protective eyewear is needed. The laser operates on a wavelength of 635 nanometers and can have an extended range up to 250 feet.

3
CHAPTER

Minor Repairs and Remodeling

PLANNING IS AN IMPORTANT PART OF ANY job. The job of remodeling is no exception. It takes a plan to get the job done correctly. The plan must have all the details worked out. This will save money, time, and effort. The work will go smoothly if the bugs have been worked out before the job is started.

You will learn

- How to diagnose problems
- How to identify needed maintenance jobs
- How to make minor repairs
- How to do some minor remodeling jobs

PLANNING THE JOB

Maintenance means keeping something operating properly. It means taking time to make sure a piece of equipment will operate tomorrow. It means doing certain things to keep a house in good repair. Many types of jobs present themselves when it comes to maintenance. The carpenter is the person most commonly called upon to do maintenance. This may range from the replacement of a lock to complete replacement of a window or door.

Remodeling means just what the word says. It means changing the looks and the function of a house. It might mean you have to put in new kitchen cabinets. The windows might need a different type of opening. Perhaps the floor is old and needs a new covering. The basement might require new paneling or tiled floors.

Working on a house when it is new calls for a carpenter who can saw, measure, and nail things in the proper place. Working on a house after it is built calls for many types of operations. The carpenter may be called upon to do a number of different things related to the trade.

Diagnosing Problems

A person who works in maintenance or remodeling needs to know what the problem actually is. That person must be able to find out what causes a problem. The next step is, of course, to decide what to do to correct the problem.

If you know how a house is built, you should be able to repair it. This means you know what has to be done to properly construct a wall or repair it if it is damaged. If the roof leaks, you need to know where and how to fix it.

In other words, you need to be able to diagnose problems in any building. Since we are concerned with the residential types here, it is important to know how things go wrong in a well-built home. In addition, it is important to be able to repair them.

Before you can remodel a house, or add on to it, you need to know how the original was built. You need to know what type of foundation was used. Can it support another story, or can the soil support what you have in mind? Are you prepared for the electrical loads? What about the sewage? How does all this fit into the addition plans? What type of consideration have you given the plumbing, drainage, and other problems?

Identifying Needed Operations

If peeling paint needs to be removed and the wall repainted, can you identify what caused the problem? This will be important later when you choose another paint. Figure 3-1 shows what can happen with paint.

Paint problems by and large are caused by the presence of moisture. The moisture may be in the wood when it is painted. Or it may have seeped in later. Take a look at Fig. 3-1 to see just what causes cracking and alligatoring. Also notice the causes of peeling, flaking, nailhead stains, and blistering.

Normal daily activity in the home of a family of four can put as much as 50 pints of water vapor in the air in one day. Since moisture vapor always seeks an area of lesser pressure, the moisture inside the house tries to become the same as the outside pressure. This equalization process results in moisture passing through walls and ceilings, sheathing, door and window casings, and the roof. The eventual result is paint damage. This occurs as the moisture passes through the exterior paint film.

Sequencing Work to Be Done

In making repairs or doing remodeling, it is necessary to schedule the work properly. It is necessary to make sure things are done in order. For example, it is difficult to paint if there is no wall to paint. This might sound ridiculous, but it is no more so than some other problems associated with getting a job done.

It is hard to nail boards onto a house if there are no nails. Somewhere in the planning you need to make sure nails are available when you need them. If things are not properly planned—in sequence—you could be ready to place a roof on the house and not have the proper size nails to do the job. You wouldn't try to place a carpet on bare joists. There must be a floor or subflooring first. This means there is some sequence that must be followed before you do a job or even get started with it.

Make a checklist to be sure you have all the materials you need to do a job *before* you get started. If there is the possibility that something won't arrive when needed, try to schedule something else so the operation can go on. Then you can pick up the missing parts later

FLAKING

CONDITION:

Siding alternatively swells and shrinks as moisture behind it is absorbed and then dries out. Paint film cracks from swelling and shrinking and flakes away from surface.

CORRECTION:

- All moisture problems must be corrected before surface is re-painted.
- Scrape and sand all peeling paint to bare wood including several inches around damaged areas. Feather edges.
- Apply primer according to label directions.
- Apply topcoat according to label directions.

BLISTERING

CONDITION:

Blistering is actually the first stage of the peeling process. It is caused by moisture attempting to escape through the existing paint film, lifting the paint away from the surface.

CORRECTION:

- All moisture problems must be corrected before repainting.
- Scrape and sand all blistering paint to bare wood several inches around blistered area.
- Feather or smooth the rough edges of the old paint by sanding.
- Apply primer according to label directions.
- Apply topcoat according to label directions.

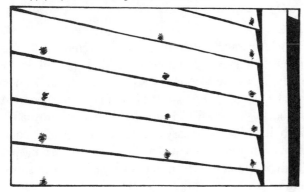

CRACKING AND ALLIGATORING

CONDITION:

Cracking and "alligatoring" are caused by (a) paint that is applied too thick, (b) too many coats of paint, (c) paint applied over a paint coat which is not completely dry, or (d) an improper primer.

CORRECTION:

- Removal of entire checked or alligatored surface may be necessary.
- Scrape and sand down the surface until smooth. Feather edges.
- Apply primer. Follow label directions.
- Apply topcoat. Follow label directions.

PEELING

CONDITION:

Peeling is caused by moisture being pulled through the paint (by the sun's heat), lifting paint away from the surface.

CORRECTION:

- All moisture problems must be corrected before surface is repainted.
- Remove all peeling and flaking paint.
- Scrape and sand all peeling paint to bare wood including several inches around damaged areas. Feather edges.
- Apply primer according to directions on label.
- Apply topcoat according to directions.

NAILHEAD STAINS

CONDITION:

Nailhead stains are caused by moisture rusting old or uncoated nails.

CORRECTION:

- All moisture problems must be corrected before repainting.
- Sand or wire brush stained paint and remove rust down to bright metal of nailhead.
- Countersink nail if necessary.
- Apply primer to nailheads. Allow to dry.
- Caulk nail holes. Allow to dry. Sand smooth.
- Apply primer to surface, following label directions.
- Apply topcoat according to label directions.

Fig. 3-1 *Causes of paint problems on houses.* (Grossman Lumber)

when it becomes available. This means that sequencing has to take into consideration the problems of supply and delivery of materials. The person who coordinates this is very important. This person can make the difference between the job being a profitable one or a money loser.

MINOR REPAIRS AND REMODELING

When doors are installed, they should fit properly. That means the closed door should fit tightly against the door stop. Figure 3-2 shows how the door should fit. If it doesn't fit properly, adjust the strike jamb side of the frame in or out. Do this until the door meets the weather stripping evenly from top to bottom.

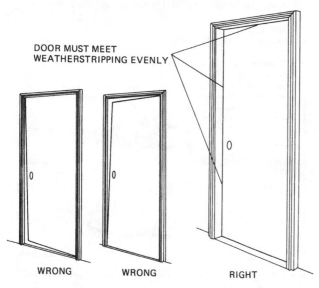

Fig. 3-2 Right and wrong ways of nailing the strike jamb. *(General Products)*

Adjusting Doors

The strike jamb can be shimmed as the hinge jamb can. Place one set of shims behind the strike plate mounting location. Renail the jamb so that it fits properly. If it is an interior door, there will be no weather stripping. Figure 3-3 shows how to shim the hinges to make sure

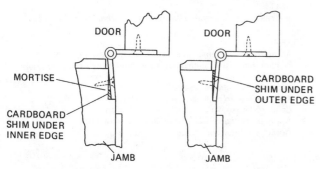

Fig. 3-3 Hinge adjustment for incorrectly fitting doors. Note the shim placement.

the door fits snugly. In some cases you might have to remove some of the wood on the door where the hinge is attached. See Fig. 3-4. This shows how the wood is removed with a chisel. You must be careful not to remove the back part of the door along the outside of the door.

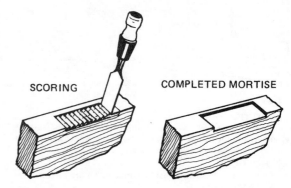

Fig. 3-4 Removing extra wood from a door. Be careful not to remove too much. *(Grossman Lumber)*

If the door binds or does not fit properly, you might have to remove some of the lock edge of the door. Bevel it as shown in Fig. 3-5 so that it fits. An old carpenter's trick is to make sure the thickness of an 8*d* nail is allowed all around the door. This usually allows enough space for the door to swell some in humid weather and not too much space when winter heat in the house dries out the wood in the door and causes it to shrink slightly.

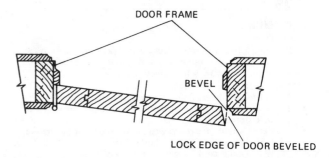

Fig. 3-5 Beveling the lock stile of a door. *(Grossman Lumber)*

Occasionally it is necessary to remove the door from its hinges by removing the hinge pins. Take the door to a vise or workbench so that the edge of the door can be planed down to fit the opening. If the amount of wood to be removed from the door is more than $1/4$ inch, it will be necessary to trim both edges of the door equally to one-half the width of the wood to be removed. Trim off the wood with a smooth or jack plane as shown in Fig. 3-6.

Sometimes it is necessary to remove the doors from their hinges and cut off the bottom because the doors were not cut to fit a room where there is carpeting. If

Fig. 3-6 *Trimming the width of a door.* (Grossman Lumber)

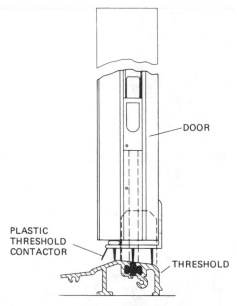

Fig. 3-7 *Adding a piece of plastic to a threshold to make sure the bottom of the door fits snugly. This keeps rain from entering the room as well as keeping out cold air in the winter.* (General Products)

this occurs, remove the door carefully from its hinges. Mark off the amount of wood that must be removed from the bottom of the door. Place a piece of masking tape over the area to be cut. Redraw your line on the masking tape, in the middle of the tape. Tape the other side of the door at the same distance from the bottom. Set the saw to cut the thickness of the door. Cut the door with a power saw or handsaw. Cutting through the tape will hold the finish on the wood. You will not have a door with splinters all along the cut edge. If the door is cut without tape, it may have splinters. This can look very bad if the door has been prefinished.

If the top of the door binds, it should be beveled slightly toward the stop. This can let it open and close more easily.

In some cases the outside door does not meet the threshold properly. It may be necessary to obtain a thicker threshold or a piece of plastic to fit on the bottom of the door as in Fig. 3-7.

Adjusting Locks

Installing a lock can be as easy as following instructions. Each manufacturer furnishes instructions with each new lock. However, in some cases you have to replace one that is around the house and no instructions can be found. Take a look at Fig. 3-8 for a step-by-step method of placing a lock into a door. Note that the lock set is typical. It can be replaced with just a screwdriver. There are 18 other brands that can be replaced by this particular lock set.

Various lock sets are available to fit the holes that already exist in a door. Strikes come in a variety of shapes too. You should choose one that fits the already grooved doorjamb.

A deadbolt type of lock is called for in some areas. This is where the crime rate is such that a more secure door is needed.

Figure 3-9 shows a lock set that is added to the existing lock in the door. This one is key-operated. This means you must have two keys to enter the door. It is very difficult for a burglar to cause this type of lock bolt to retract.

In some cases it is desired that once the door is closed, it is locked. This can be both an advantage and a disadvantage. If you go out without your key, you are in trouble. This is especially true if no one else is home. However, it is nice for those who are a bit absent-minded and forget to lock the door once it is closed. The door locks automatically once the door is pushed closed.

Figure 3-10 shows how to insert a different type of lock, designed to fit into the existing drilled holes. All you need is a screwdriver to install it. Figure 3-11 shows how secure the lock can be with double cylinders to fit through the holes on the doorjamb.

About the only trouble with a lock set is the loosening of the doorknob. It can be tightened with a screwdriver in most cases. If the lock set has two screws near the knob on the inside of the door, simply align the lock set and tighten the screws. If it is another type—with no screws visible—just release the small tab that sticks up inside the lock set; it can be seen through the brass plate around the knob close to the wooden part of the door. This will allow the brass plate or ring to be rotated and removed. Then it is a matter of tightening the screws found inside the lock set. Tighten the screws and replace the cover plate.

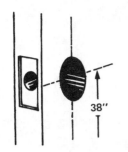 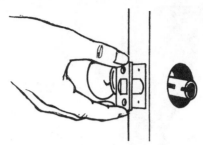 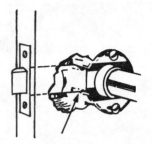

STEP 1: Prepare door; drill for bolt and lock mechanism.

STEP 2: Insert bolt and lock mechanism.

STEP 3: Engage bolt and lock mechanism; fasten face plate.

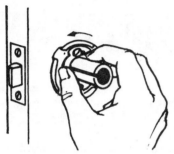 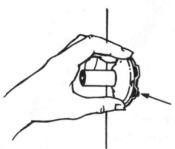 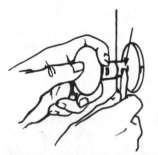 

STEP 4: Put on clamp plate; Tighten screws.

STEP 5: "Snap-On" rose.

 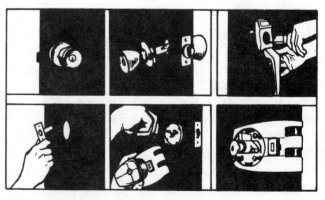

STEP 6: Apply knob on spindle by depressing spring retainer.

STEP 7: Mortise for latch bolt; fasten strike with screws.

Fig. 3-8 *Fitting a lock set into a door.* (National Lock)

Fig. 3-9 *Key-operated auxiliary lock.* (Weiser Lock)

Fig. 3-10 *Putting in a lock set.* (Weiser Lock)

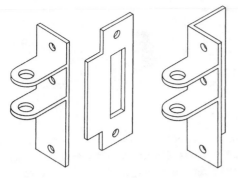

Fig. 3-11 *Strikes for single- and double-cylinder locks.*

The strike plate may work loose or the door may settle slightly. This means the striker will not align with the strike plate. Adjust the plate or strike screws if necessary. In some cases it might be easier to remove the strike plate and file out a small portion to allow the bolt to fit into the hole in the doorjamb.

If the doorknob becomes green or off-color, remove it. Polish the brass and recoat the knob with a covering of lacquer and replace it. In some cases the brass is plated and will be removed with the buffing. Here you may want to add a favorite shade of good metal enamel to the knob and replace it. This discoloring does occur in bathrooms where the moisture attacks the doorknob lacquer coating.

Installing Drapery Hardware

One often overlooked area of house building is the drapery hardware. People who buy a new house are faced with the question What do I do to make these windows attractive? Installing window hardware can be a job in itself. Installing conventional adjustable traverse rods is a task that can be easily done by the carpenter, in some cases, or the homeowner. There are specialists who do these things. However, it is usually an extra service that the carpenter gets paid for after the house is finished and turned over to the home owner.

Decorative traverse rods are preferred by those who like period furniture. See Fig. 3-12. Installation of a traverse rod is shown step by step in Figs. 3-13 through 3-17.

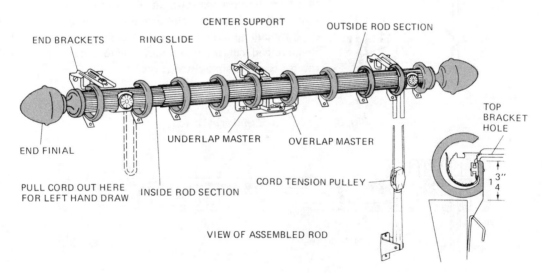

Fig. 3-12 *An adjustable decorative traverse rod.*

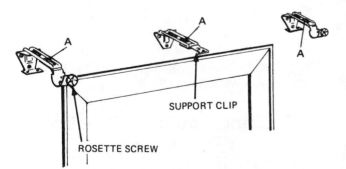

Install end brackets above and 6" to 18" to the sides of casing. When installing on plaster or wallboard and screws do not anchor to studding, plastic anchors or other installation aids may be needed to hold brackets securely. Place center supports (provided with longer size rods only) equidistant between end brackets. Adjust bracket and support projection at screws. "A." *NOTE: Attach rosette screws to end brackets if not already assembled.*

Fig. 3-13 *Install end brackets and center support.*

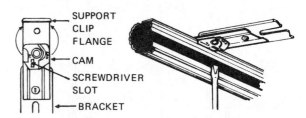

Place end finials on rod. Finial with smaller diameter shaft fits in "inside" rod section, larger one fits in "outside" section. Place last ring between bracket and end finial.

Extend rod to fit brackets. Place rod in brackets so tongue on bottom of bracket socket fits into hole near end of rod. Tighten rosette screw.

When center supports are used, insert top of rod in support clip flange. If outside rod section, turn cam with screwdriver counterclockwise to lock in rod. If inside rod section, turn cam clockwise.

Fig. 3-14 *Place rod in bracket.*

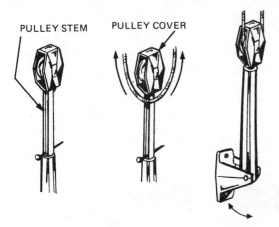

Right-hand draw is standard. If left hand draw is desired, simply pull cord loop down from pulley wheels on left side of rod.

Fasten tension pulley to sill, wall or floor at point where cord loop falls from traverse rod. Lift pulley stem and slip nail through hole in stem. Pull cord up through bottom of pulley cover. Take up excess slack by pulling out knotted cord from back of overlap master slide. Re-knot and cut off extra cord. Remove nail. Pulley head may be rotated to eliminate twisted cords.

Fig. 3-15 *Install cord tension pulleys.*

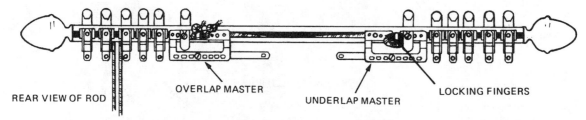

Pull draw cord to move overlap master to end of rod. Holding cords taut, slide underlap master to opposite end of rod. Then fasten cord around locking fingers on underlap master to stop underlap master from slipping. Then center both masters.

Fig. 3-16 *Adjusting master slides.*

Figure 3-18 shows some of the regular-duty types of corner and bay window drapery rods. Note the various shapes of bay windows that are made. The rods are made to fit the bay windows.

Figure 3-19 shows more variations of bay windows. These call for some interesting rods. Of course, in some cases the person looks at the cost of the rods and hardware and decides to put one straight rod across the bay and not follow the shape of the windows.

This can be the least expensive, but it negates the effect of having a bay window in the first place.

Repairing Damaged Sheetrock Walls (Drywall)

In drywall construction the first areas to show problems are over joints or fastener heads. Improper application of either the board or the joint treatment may be at

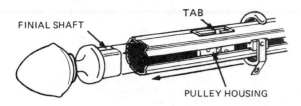

FINIAL SHAFT

TAB

PULLEY HOUSING

Slip off end finial. Remove extra ring slides by pulling back on tab and sliding rings off end of rod. Be sure to leave the last ring between pulley housing and end finial.

To train draperies, "break" fabric between pleats by folding the material toward the window. Fold pleats together and smooth out folds down the entire length of the draperies. Tie draperies with light cord or cloth. Leave tied two or three days before operating.

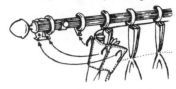

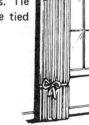

Fig. 3-17 *(A) Remove unused ring slides; (B) training draperies.*

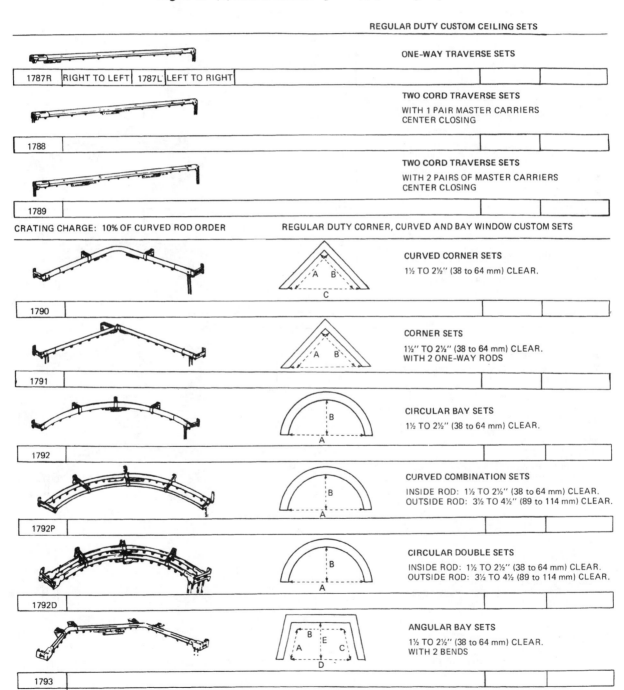

REGULAR DUTY CUSTOM CEILING SETS

ONE-WAY TRAVERSE SETS

| 1787R | RIGHT TO LEFT | 1787L | LEFT TO RIGHT | | | |

TWO CORD TRAVERSE SETS
WITH 1 PAIR MASTER CARRIERS CENTER CLOSING

| 1788 | | | |

TWO CORD TRAVERSE SETS
WITH 2 PAIRS OF MASTER CARRIERS CENTER CLOSING

| 1789 | | | |

CRATING CHARGE: 10% OF CURVED ROD ORDER

REGULAR DUTY CORNER, CURVED AND BAY WINDOW CUSTOM SETS

CURVED CORNER SETS
1½ TO 2½'' (38 to 64 mm) CLEAR.

| 1790 | | | |

CORNER SETS
1½'' TO 2½'' (38 to 64 mm) CLEAR.
WITH 2 ONE-WAY RODS

| 1791 | | | |

CIRCULAR BAY SETS
1½ TO 2½'' (38 to 64 mm) CLEAR.

| 1792 | | | |

CURVED COMBINATION SETS
INSIDE ROD: 1½ TO 2½'' (38 to 64 mm) CLEAR.
OUTSIDE ROD: 3½ TO 4½'' (89 to 114 mm) CLEAR.

| 1792P | | | |

CIRCULAR DOUBLE SETS
INSIDE ROD: 1½ TO 2½'' (38 to 64 mm) CLEAR.
OUTSIDE ROD: 3½ TO 4½'' (89 to 114 mm) CLEAR.

| 1792D | | | |

ANGULAR BAY SETS
1½ TO 2½'' (38 to 64 mm) CLEAR.
WITH 2 BENDS

| 1793 | | | |

Fig. 3-18 *Regular-duty custom ceiling sets of traverse rods; regular-duty corner, curved, and bay window custom sets.* (Kenny)

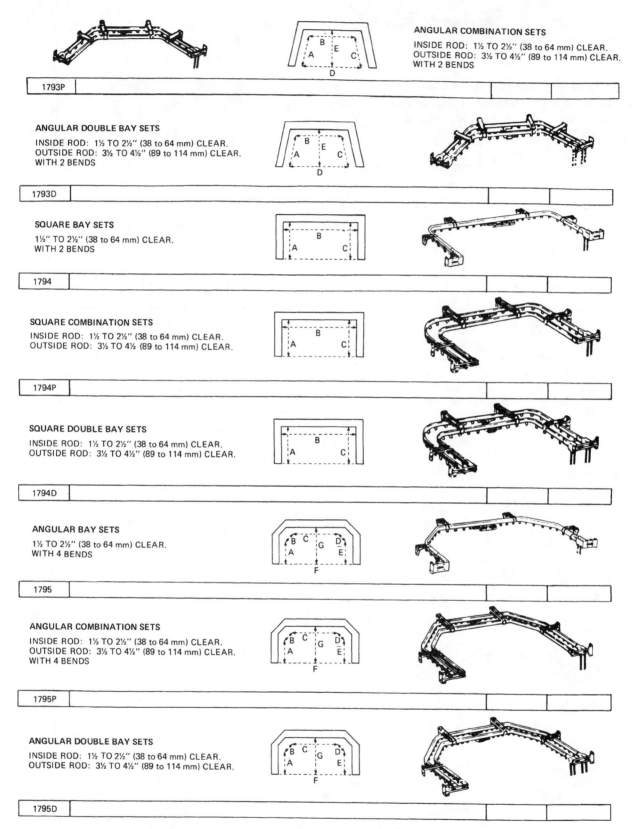

ANGULAR COMBINATION SETS

INSIDE ROD: 1½ TO 2½" (38 to 64 mm) CLEAR.
OUTSIDE ROD: 3½ TO 4½" (89 to 114 mm) CLEAR.
WITH 2 BENDS

1793P

ANGULAR DOUBLE BAY SETS

INSIDE ROD: 1½ TO 2½" (38 to 64 mm) CLEAR.
OUTSIDE ROD: 3½ TO 4½" (89 to 114 mm) CLEAR.
WITH 2 BENDS

1793D

SQUARE BAY SETS

1½" TO 2½" (38 to 64 mm) CLEAR.
WITH 2 BENDS

1794

SQUARE COMBINATION SETS

INSIDE ROD: 1½ TO 2½" (38 to 64 mm) CLEAR.
OUTSIDE ROD: 3½ TO 4½ (89 to 114 mm) CLEAR.

1794P

SQUARE DOUBLE BAY SETS

INSIDE ROD: 1½ TO 2½" (38 to 64 mm) CLEAR.
OUTSIDE ROD: 3½ TO 4½" (89 to 114 mm) CLEAR.

1794D

ANGULAR BAY SETS

1½ TO 2½" (38 to 64 mm) CLEAR.
WITH 4 BENDS

1795

ANGULAR COMBINATION SETS

INSIDE ROD: 1½ TO 2½" (38 to 64 mm) CLEAR.
OUTSIDE ROD: 3½ TO 4½" (89 to 114 mm) CLEAR.
WITH 4 BENDS

1795P

ANGULAR DOUBLE BAY SETS

INSIDE ROD: 1½ TO 2½" (38 to 64 mm) CLEAR.
OUTSIDE ROD: 3½ TO 4½" (89 to 114 mm) CLEAR.

1795D

Fig. 3-19 *More regular-duty corner, curved, and bay window custom rods.* (Kenny)

fault. Other conditions existing on the job can also be responsible for reducing the quality of the finished gypsum board surface.

Panels improperly fitted

Cause   Forcibly wedging an oversize panel into place is the cause. This bows the panel and builds in stresses. The stress keeps it from contacting the framing. See Fig. 3-20.

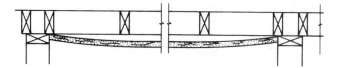

Fig. 3-20   *Forcibly fitted piece of gypsum board.* (U.S. Gypsum)

Result   After nailing, a high percentage of the nails on the central studs probably will puncture the paper. This may also cause joint deformation,

Remedy   Remove the panel. Cut it to fit properly. Replace it. Fasten from the center of the panel toward the ends and edges. Apply pressure to hold the panel tightly against the framing while you drive the fasteners.

Panels with damaged edges

Cause   Paper-bound edges have been damaged or abused. This may result in ply separation along the edge. Or it may loosen the paper from the gypsum core. Or it may fracture or powder the core itself. Damaged edges are more susceptible to ridging after joint treatment.

Remedy   Cut back any severely damaged edges to the sound board before application.

Prevention   Avoid using board with damaged edges that may easily compress. Damaged edges can take on moisture and swell. Handle sheetrock with care.

Panels loosely fastened

Cause   Framing members are uneven because of misalignment or warping. If there is lack of hand pressure on the panel during fastening, loosely fitting panels can result. See Fig. 3-21.

Fig. 3-21   *Damaged edges and their effect on a joint.* (U.S. Gypsum)

Remedy   When panels are fastened with nails, during final blows of the hammer use your hand to apply additional pressure to the panel adjacent to the nail. See Fig. 3-22.

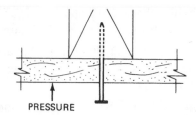

Fig. 3-22   *Apply hand pressure (see arrow) while nailing the board to the stud.* (U.S. Gypsum)

Prevention   Correct framing imperfections before you apply the panels. Use screws or adhesive method instead of nails.

Surface fractured after application

Cause   Heavy blows or other abuse have fractured finished wall surface. If the break is too large to repair with joint compound, do the following.

Remedy   In the shape of an equilateral triangle around the damaged area, remove a plug of gypsum. Use a keyhole saw. Slope the edges 45°. Cut a corresponding plug from a sound piece of gypsum. Sand the edges to an exact fit. If necessary, cement an extra slat of gypsum panel to the back of the face layer to serve as a brace. Butter the edges and finish as a butt joint with joint compound.

Framing members out of alignment

Cause   Because of misaligned top plate and stud, hammering at points *X* in Fig. 3-23 as panels are applied on both sides of the partition will probably result in nail heads puncturing the paper or cracking the board. If framing members are more than ¼ inch out of

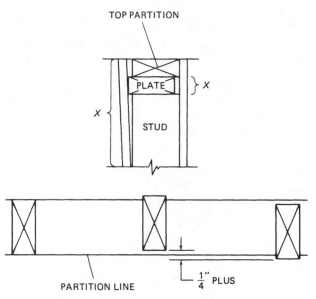

Fig. 3-23   *If framing members are bowed or misaligned, shims are needed if the wallboard is to fit properly.* (U.S. Gypsum)

alignment with adjacent members, it is difficult to bring panels into firm contact with all nailing surfaces.

Remedy　Remove or drive in problem fasteners and drive new fasteners only into members in solid contact with the board.

Prevention　Check the alignment of studs, joists, headers, blocking, and plates before you apply panels. Correct before proceeding. Straighten badly bowed or crowned members. Shim out flush with adjoining surfaces. Use adhesive attachment.

Members twisted

Cause　Framing members have not been properly squared with the plates. This gives an angular nailing surface. See Fig. 3-24. When panels are applied, there is a danger of fastener heads puncturing the paper or of reverse twisting of a member as it dries out. This loosens the boards and can cause fastener pops.

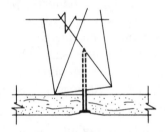

Fig. 3-24　*Framing members are improperly squared. (U.S. Gypsum)*

Remedy　Allow the moisture content in the framing to stabilize. Remove the problem fasteners. Refasten with carefully driven screws.

Prevention　Align all the twisted framing members before you apply the board.

Framing protrusions

Cause　Bridging, headers, fire stops, or mechanical lines have been installed improperly. See Fig. 3-25. They may project out past the face of the framing member. This prevents the board or drywall surface from meeting the nailing surface. The result can be a loose board. Fasteners driven in this area of protrusion will probably puncture the face paper.

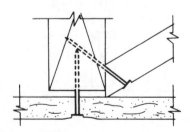

Fig. 3-25　*Nail head goes through but doesn't pull the board up tightly against the stud. The stud is prevented from meeting the nailing surface by a piece of bridging out of place. (U.S. Gypsum)*

Remedy　Allow the moisture content in the framing to stablilize. Remove the problem fasteners. Realign the bridging or whatever is out of alignment. Refasten with carefully driven screws.

Puncturing of face paper

Cause　Poorly formed nail heads, careless nailing, excessively dry face paper, or a soft core can cause the face paper to puncture. Nail heads that puncture the paper and shatter the core of the panel are shown in Fig. 3-26. They have little grip on the board.

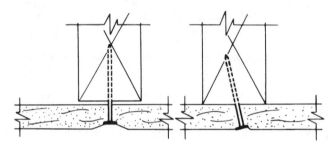

Fig. 3-26　*Puncturing of the face paper. (U.S. Gypsum)*

Remedy　Remove the improperly driven fastener. Properly drive a new fastener.

Prevention　Correcting faulty framing and driving nails properly produce a tight attachment. There should be a slight uniform dimple. See Fig. 3-27 for the proper installation of the fastener. A nail head bears on the paper. It holds the panel securely against the framing member. If the face paper becomes dry and brittle, its low moisture content may aggravate the nail cutting. Raise the moisture content of the board and the humidity in the work area.

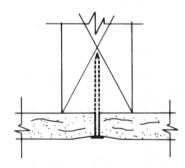

Fig. 3-27　*Proper dimple made with the nail head into the drywall board. (U.S. Gypsum)*

Nail pops from lumber shrinkage

Cause　Improper application, lumber shrinkage, or a combination of the two is the problem. With panels held reasonably tight against the framing member and with proper-length nails, normally only severe shrinkage of the lumber will cause nail pops. But if the panels are nailed loosely, any inward pressure on the panel will push the nail head through its thin covering pad of compound. Pops resulting from "nail creep" occur when shrinkage of the wood framing exposes nail shanks and consequently loosens the panels. See Fig. 3-28.

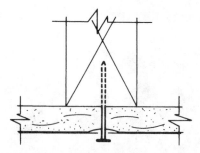

Fig. 3-28 *Popped nail head.* (U.S. Gypsum)

Remedy Repairs usually are necessary only for pops that protrude 0.005 inch or more from the face of the board. See Fig. 3-28. Smaller protrusions may need to be repaired if they occur in a smooth gloss surface or flat-painted surface under extreme lighting conditions. Those that appear before or during decoration should be repaired immediately. Pops that occur after one month's heating or more are usually caused wholly or partly by wood shrinkage. They should not be repaired until near the end of the heating season. Drive the proper nail or screw about 1½ inches from the popped nail while applying sufficient pressure adjacent to the nail head to bring the panel in firm contact with the framing. Strike the popped nail lightly to seat it below the surface of the board. Remove the loose compound. Apply finish coats of compound and paint.

These are but a few of the possible problems with gypsum drywall or sheetrock. Others are cracking, surface defects, water damage, and discoloration. All can be repaired with the proper tools and equipment. A little skill can be developed over time. However, in most instances it is necessary to redecorate a wall or ceiling. This can become a problem of greater proportions. It is best to make sure the job is done correctly the first time. This can be done by looking at some of the suggestions given under *Prevention.*

Installing New Countertops

Countertops are usually covered by a plastic laminate. Formica is usually applied to protect the wooden surface and make it easier to clean. (However, Formica is only one of the trademarks for plastic laminates.) Plastic laminates are easy to work with since they can be

Fig. 3-29 *Cutting plastic laminate with a saber saw.* (U.S. Gypsum)

cut with either a power saw or a handsaw. Just be sure to cut the plastic laminate face down when you use a portable electric saw. This minimizes chipping. See Fig. 3-29.

If you use a handsaw, as in Fig. 3-30, make sure you use a low angle and cut only on the downward stroke.

Fig. 3-30 *Cutting plastic laminate with a handsaw.* (Grossman Lumber)

Before applying the laminate to the surface of the counter, you have to coat both the laminate and the counter with adhesive. See Fig. 3-31A. This is usually a contact cement. That means both surfaces must be dry to the touch before they are placed in contact with each other. See Fig. 3-31B and C. Make sure the cement is given at least 15 minutes and no more than 1 hour for drying time. See Fig. 3-32. If the cement sinks into the work surface, it may be best to apply a second coat. See Fig. 3-33A and B. Make sure it is dry before you apply the laminate to the work surface of the counter.

To get a perfect fit, you might have to place a piece of brown paper on the work surface. Then slide the plastic over the paper. Slip the paper from under the plastic laminate slowly. As the paper is removed, the two adhesive surfaces will contact and stick. Don't let the

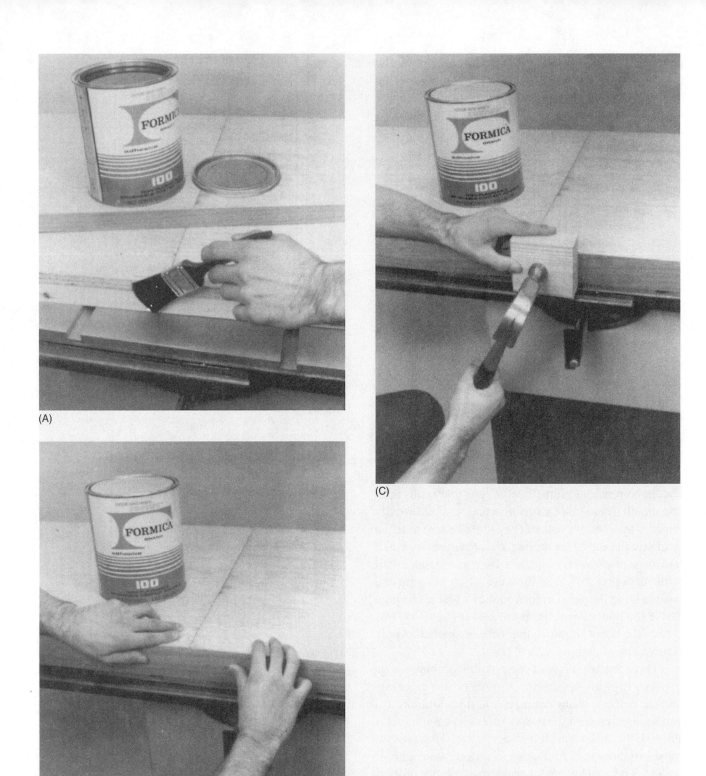

Fig. 3-31 *Applying the strip of laminate to the edge. (A) Flip the top right side up. Apply contact cement to laminate and core self-edge* (Formica). *(B) Bond the strip of laminate to the edge, using your fingertips to keep the surfaces apart as you go* (Formica). *(C) Apply pressure immediately after bonding by using a hammer and a clean block of hardwood.* (Formica)

Fig. 3-32 *Placing the glue-coated plastic laminate on the bottom over a piece of paper.* (Grossman Lumber)

cemented surfaces touch until they are in the proper location. Figure 3-33C and D shows another method of applying the laminate.

Once the paper has been removed and the surfaces are aligned, apply pressure over the entire area. Carefully pound a block of wood with a hammer on the laminated top. This bonds the laminate to the surface below. See Fig. 3-33E for another method.

Trimming edges Use a router as shown in Fig. 3-34 to finish up the edges. A saber saw or hand plane can be used if you do not have a router. The edge of the plywood that shows can be covered with plastic laminate, wood molding, or metal strips made for the job. Figure 3-35

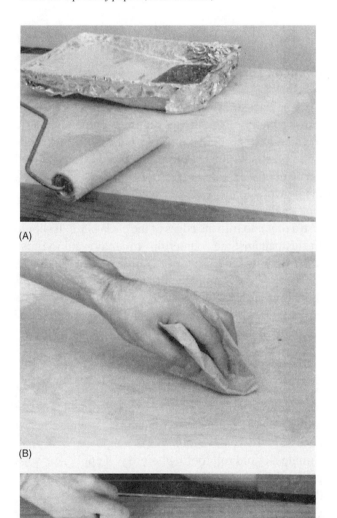

(A)

(B)

(C)

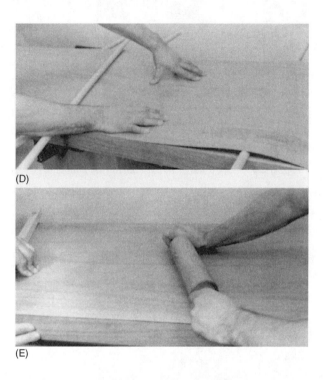

(D)

(E)

Fig. 3-33 *(A) Use a paint roller and work from a tray to apply the contact cement. Lining the tray with aluminum foil speeds cleanup later* (Formica)*. (B) Surfaces are ready for bonding when the glue does not adhere to clean kraft paper* (Formica)*. (C) Align the laminate over the core. As you remove the $^1/_4$-inch sticks, the laminate is bonded to the wood* (Formica)*. (D) Dowel rods can also be used to prevent bonding until they are removed* (Formica)*. (E) Use an ordinary rolling pin to apply pressure to create a good bond.* (Formica)

Fig. 3-34 *Using a router to edge the plastic laminate.* (Formica)

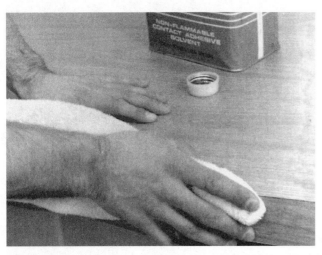

Fig. 3-35 *Wipe the surfaces clean with a rag and thinner. Use the thinner very sparingly, since it can cause delamination at the edges.* (Formica)

shows the cleanup operation after the laminate has been applied.

Fitting the sink in the countertop is another step in making a finished kitchen. You will need a clamp-type sink rim installation kit. Apply the kit to your sink to be sure it is the correct size. Then cut the countertop hole about ¹⁄₈ inch larger all around than the rim. Make the cutout with a keyhole saw, a router, or a saber saw. See Fig. 3-36.

Fig. 3-36 *Marking around the sink rim for the cutout in the countertop.* (Grossman Lumber)

Start the cutout by drilling a row of holes and leaving ¹⁄₁₆ inch to accommodate the leg of the rim. Install the sink and rim. Or, you can go on to the backsplash.

The backsplash is a board made from ³⁄₄-inch plywood which is nailed to the wall. It is perpendicular to the countertop. Finish off the backsplash with the same plastic laminate before you mount it to the wall permanently. In some cases it may be easier to mount the plywood to the wall with an adhesive, since the drywall is already in place at this step in the construction. Use a router to trim the edges of the backsplash. It should fit flush against the countertop. You may want to finish it off with the end grain of the plywood covered with the same plastic laminate or with metal trim.

Figure 3-37 shows how the sink is installed in the countertop. Also note the repair method used to remove a bubble in the plastic laminate.

Repairing a Leaking Roof

In most areas when a reroofing job is under consideration, a choice must be made between removing the old roofing or permitting it to remain. It is generally not necessary to remove old wood singles, old asphalt shingles, or old roll roofing before you apply a new asphalt roof—that is, if a competent inspection indicates that

1. The existing deck framing is strong enough to support the weight of workers and the additional new roofing. This means it should also be able to support the usual snow and wind loads.

2. The existing deck is sound and will provide good anchorage for the nails used in applying the new roofing.

Old roofing to stay in place If the inspection indicates that the old wood shingles may remain, the surface

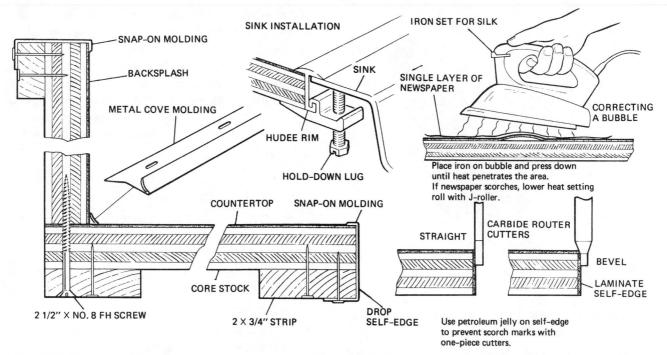

Fig. 3-37 *Details of installing a sink in a laminated countertop. Note the method used to remove a bubble caught under the laminate.* (Formica)

of the roof should be carefully prepared to receive the new roofing.

This may be done as follows:

1. Remove all loose or protruding nails, and renail the shingles in a new location.
2. Nail down all loose shingles.
3. Split all badly curled or warped old shingles and nail down the segments.
4. Replace missing shingles with new ones.
5. When shingles and trim at the eaves and rakes are badly weathered, and when the work is being done in a location subject to the impact of unusually high winds, the shingles at the eaves and rakes should be cut back far enough to allow for the application, at these points, of 4- to 6-inch wood strips, nominally 1 inch thick. Nail the strips firmly in place, with their outside edges projecting beyond the edges of the deck the same distance as did the wood shingles. See Fig. 3-38.
6. To provide a smooth deck to receive the asphalt roofing, it is recommended that beveled wood "feathering" strips be used along the butts of each course of old shingles.

Old roofing (asphalt) shingles to remain in place
If the old asphalt shingles are to remain in place, nail down or cut away all loose, curled, or lifted shingles. Remove all loose and protruding nails. Remove all badly worn edging strips and replace with new. Just before you

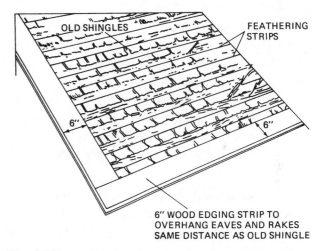

Fig. 3-38 *Treatment of rakes and eaves when reroofing in windy locations.* (Bird and Son)

apply the new roofing, sweep the surface clean of all loose debris.

Square-butt strip shingles to be recovered with self-sealing square-butt strip shingles The following application procedure is suggested to minimize the uneven appearance of the new roof. All dimensions are given assuming that the existing roof has been installed with the customary 5-inch shingle exposure.

Starter Course Cut off the tabs of the new shingle, using the head portion equal in width to the exposure of the old shingle. This is normally 5 inches for the starter shingle. See Fig. 3-39.

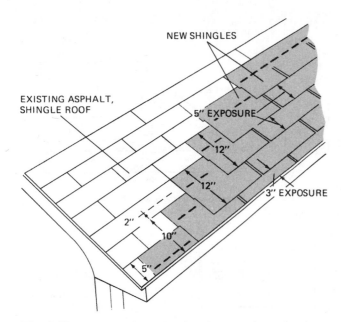

Fig. 3-39 *Exposure of new shingles when reroofing.* (Bird and Son)

First Course Cut 2 inches from the top edge of a full-width new shingle. Align this cut edge with the butt edge of the old shingle.

Second Course Use a full-width shingle. Align the top edge with the butt edge of the old shingle in the next course. Although this will reduce the exposure of the first course, the appearance should not be objectionable, as this area is usually concealed by the gutter.

Third Course and All Others Use full-width shingles. Align the top edges with the butts of the old shingles. Exposure will be automatic and will coincide with that of the old roof.

Old lock-down or staple-down shingles These shingles should be removed before reroofing. They have an uneven surface, and the new shingles will tend to conform to it. If a smoother-surface base is designed, the deck should be prepared as described in *Old Roofing to Be Removed* below.

New shingles over old roll roofing When new asphalt roofing is to be laid over old roll roofing without removing the latter, proceed as follows to prepare the deck:

1. Slit all buckles, and nail segments down smoothly.
2. Remove all loose and protruding nails.
3. If some of the old roofing has been torn away, leaving pitchy knots and excessively resinous areas exposed, cover the defects with sheet metal patches made from galvanized iron, painted tin, zinc, or copper having a thickness approximately equal to 26 gauge.

Old roofing to be removed When the framing supporting the existing deck is not strong enough to support the additional weight of roofing and workers during application, or when the decking material is so far gone that it will not furnish adequate anchorage for the new roofing nails, then the old roofing, regardless of type, must be removed before new roofing is applied. The deck should then be prepared for the new roofing as follows:

1. Repair the existing roof framing where required to level and true it up and to provide adequate strength.
2. Remove all rotted or warped old sheathing, and replace it with new sheathing of the same kind.
3. Fill in all spaces between boards with securely nailed wood strips of the same thickness as the old deck. Or, move existing sheathing together and sheath the remainder of the deck.
4. Pull out all protruding nails and renail sheathing firmly at new nail locations.
5. Cover all large cracks, slivers, knot holes, loose knots, pitchy knots, and excessively resinous areas with sheet metal securely nailed to the sheathing.
6. Just before you apply the new roofing, sweep the deck thoroughly to clean off all loose debris.

Old built-up roofs

If the deck has adequate support for nails When the pitch of the deck is below 4 inches per foot but not less than 2 inches per foot, and if the deck material is sound and can be expected to provide good nail-holding power, then any old slag, gravel, or other coarse surfacing materials should first be removed. This should leave the surface of the underlying felts smooth and clean. Apply the new asphalt shingles directly over the felts according to the manufacturer's recommendations for low-slope application.

If the deck material is defective and cannot provide adequate security All old material down to the upper surface of the deck should be removed. The existing deck material should be repaired. Make it secure to the underlying supporting members. Sweep it clean before you apply the new roofing.

Patching a roof In some cases it is not necessary to replace the entire roof to plug a leak. In most instances you can visually locate the place where the leak is occurring. There are a number of roof cements that can be used to plug the hole or cement the shingles down. In some cases it is merely a case of backed-up water. To keep this from happening, heating cables may have to be placed on the roof to melt the ice.

Replacing Guttering

Guttering comes in both 4- and 5-inch widths. The newer aluminum gutter is usually available in 5-inch widths. The aluminum has a white baked-on finish. It does not require soldering, painting, or priming. It is easily handled by one person. The light weight can have some disadvantages. Its ability to hold ice or icicles in colder climates is limited. It can be damaged by the weight of ice buildup. However, its advantages usually outweigh its disadvantages. Not only is it lighter, prefinished on its exterior, and less expensive, but also it is very easy to put up.

The aluminum type of guttering is used primarily as a replacement gutter. The old galvanized type requires a primer before it is painted. In most cases, it does not hold paint even when primed. It becomes an unsightly mess easily with extremes in weather. It is necessary to solder the galvanized guttering. These soldered points do not hold if the weather is such that it heats up and cools down quickly. The solder joints are subject to breaking or developing hairline cracks which leak. These leaks form large icicles in northern climates and put an excessive load on the nails that support the gutter. Once the nails have been worked loose by the expansion and contraction, it is only a matter of time before the guttering begins to sag. If water gets behind the gutter and against the fascia board, it can cause the board to rot. This further weakens the drainage system. The fascia board is used to support the whole system.

Figure 3-40A shows that the drainage system should be lowered at one end so that the water will run down the gutter to the downspout. About $1/4$ inch for every 10 feet is sufficient for proper drainage. Figure 3-40B shows a hidden bracket hanger. It is used to support the gutter.

In Fig. 3-40C you can see the method used to mount this concealed bracket. The rest of the illustrations in Fig. 3-40 show how the system is put together to drain water from the roof completely. Each of the tools needed when replacing guttering has a name. The tools and their names are shown in Fig. 3-41. The caulking gun comes in handy to caulk places that were left unprotected by removal of the previous system. The holes for the support of the previous system should be caulked. Newer types of cements can be used to make sure each connection of the inside and outside corners and the end caps are watertight.

Once the water leaves the downspout, it is spread onto the lawn or it is conducted through plastic pipes to the storm sewer at the curb. Some locations in the country will not allow the water to be emptied onto the lawn. The drains to the storm sewer are placed in operation when the house is built. This helps control the seepage of water back into the basement once it has been pumped out with a sump pump. The sump pump also empties into the drainage system and dumps water into the storm sewer in the street.

Replacing a Floor

Many types of floor coverings are available. You may want to check with a local dealer before deciding just which type of flooring you want. There are continuous rolls of linoleum, or there are 9-inch × 9-inch squares of soft tile. There are 12-inch × 12-inch squares of carpet that can be placed down with their own adhesive. The type of flooring chosen determines the type of installation method to use.

Staple-down floor   This type of floor is rather new. It can be placed over an old floor. A staple gun is used to fasten down the edges.

Figures 3-42 through 3-44 show the procedure required to install this type of floor. The type of flooring is so flexible it can be folded and placed in the trunk of even the smallest car. Unroll the flooring in the room and move it into position (Fig. 3-42). In the 12-foot width (it also comes in 6-foot width), it covers most rooms without a seam. In the illustration it is being laid over an existing vinyl floor. It can also be installed over plywood, particleboard, concrete, and most other subfloor materials.

To cut away excess material (see Fig. 3-43), use a metal straightedge or carpenter's square to guide the utility knife. Install the flooring with a staple every 3 inches close to the trim. (See Fig. 3-44). This way the quarter-round trim can be installed over the staples, and they will be out of sight. Cement is used in places where a staple can't penetrate. For a concrete floor, use a special adhesive around the edges. Any dealer who sells this flooring has the adhesive.

The finished job looks professional even when it is done by a do-it-yourselfer. This flooring has a built-in memory. When it was rolled face-side-out at the factory for shipment, the outer circumference of the roll was stretched. After it is installed in the home, the floor gently contracts, trying to return to the dimensions it had before it was rolled up. This causes any slack or wrinkles that might have been left in the flooring to gradually be taken up by the memory action.

No-wax floor   One of the first rooms that comes up for improvement is the kitchen. Remodeling may be a major project, but renewing a floor is fairly simple.

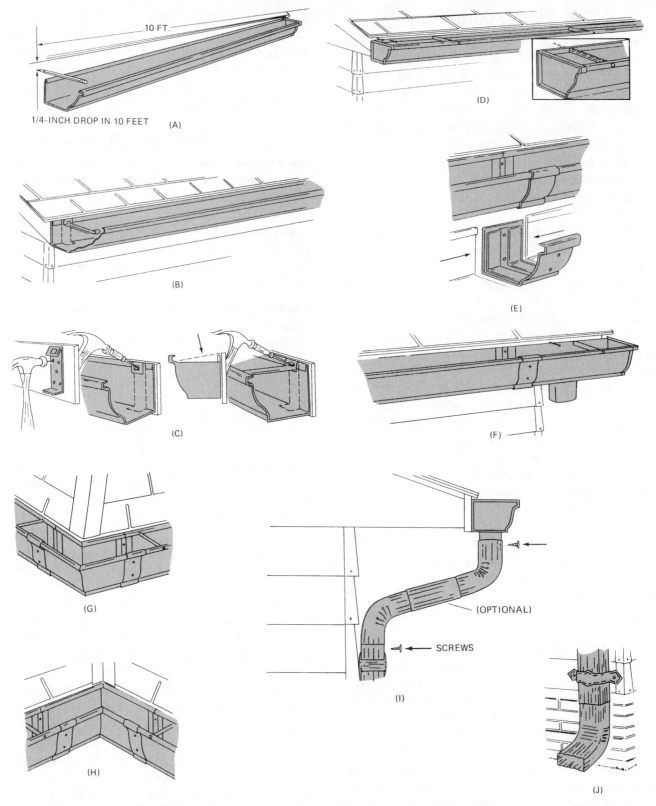

Fig. 3-40 *Replacing an existing gutter and downspout system.* (Sears, Roebuck and Co.)

No-wax flooring comes in the standard widths—6 and 12 feet. In some instances it doesn't need to be tacked or glued down. However, in most cases it should be cemented down. It fits directly over most floors, provided they are clean, smooth, and well bonded. Make sure any holes in the existing flooring are filled and smoothed over. In the case of concrete basement floors, just vacuum them or wash them thoroughly and allow them to dry.

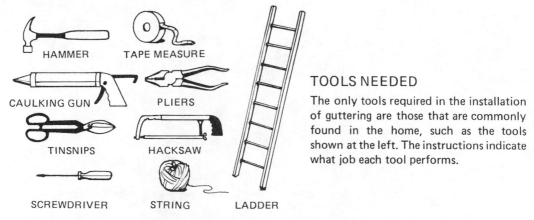

TOOLS NEEDED

The only tools required in the installation of guttering are those that are commonly found in the home, such as the tools shown at the left. The instructions indicate what job each tool performs.

Fig. 3-41 *Tools needed in the installation of guttering.*

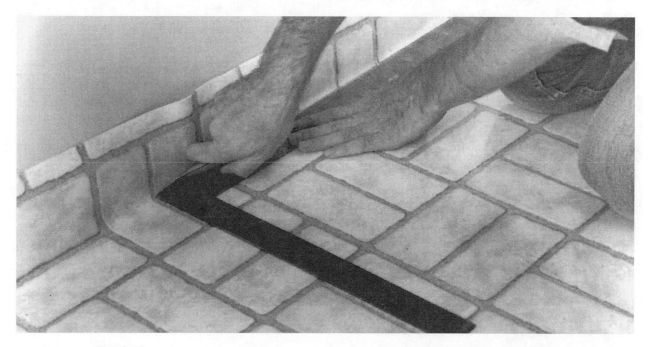

Fig. 3-42 *Unroll the flooring and allow it to sit face up overnight before installing it.* (Armstrong Cork)

Fig. 3-43 *Use a utility knife and carpenter's square to make sure the cut is straight.* (Armstrong Cork)

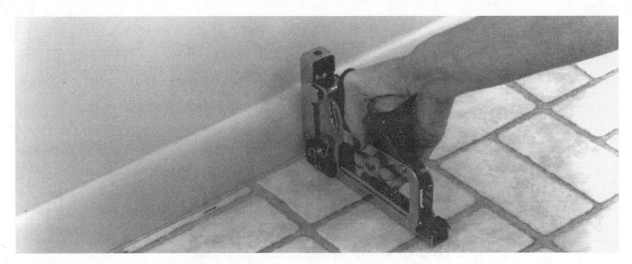

Fig. 3-44 *Place a staple every 3 inches along the kickboard. These staples will be covered by molding.* (Armstrong Cork)

The tools needed to install a new floor are a carpenter's square, chalk line, adhesive, trowel, and knife or scissors.

The key to a perfect fit is taking accurate room measurements. See Fig. 3-45. Diagram the floor plan on a chart, noting the positions of cabinets, closets, and doorways.

After you transfer the measurements from the chart to the flooring material, cut along the chalk lines, using a sharp knife and a straightedge. Transfer the measurements and cut the material in a room where the material can lie flat. Cardboard under the cut lines will protect the knife blade.

Return the material to the room where it is to be installed. Put it in place. Roll back one-half of the material. Spread the adhesive. Unroll the material onto the adhesive while it is still wet. Repeat the same steps with the rest of the material to finish the job. See Fig. 3-46.

The finished job makes any kitchen look new. All that is needed in the way of maintenance is a sponge mop with detergent.

Floor tiles Three of the most popular kinds of self-adhering tiles are vinyl-asbestos, no-wax, and vinyl tiles that contain no asbestos filler.

The benefit of no-wax tiles is obvious from the name. They have a tough, shiny, no-wax wear surface. You pay a premium price for no-wax tiles.

Vinyl tiles are not no-wax, but they are easier to clean than vinyl-asbestos tiles. Their maintenance benefits are the result of a nonporous vinyl wear surface that resists dirt, grease, and stains better than vinyl-asbestos.

Any old tile or linoleum floor can be covered with self-adhering tiles provided the old material is smooth and well bonded to the subfloor. Just make sure the surface is clean and old wax is removed.

Fig. 3-45 *Room measurements are essential to a good job.* (Armstrong Cork)

Fig. 3-46 *Spread the adhesive on half of the floor and place the flooring material down. Then do the other half.* (Armstrong Cork)

Putting down the tiles   Square off the room with a chalk line. Open the carton, and peel the protective paper off the back of the tile. See Fig. 3-47.

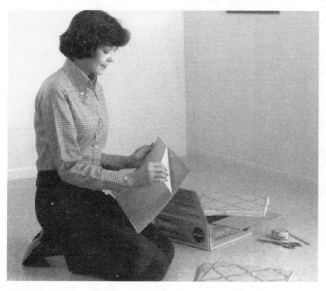

Fig. 3-47 *Peel the paper off the back of a floor tile and place the tile in a predetermined spot.* (Armstrong Cork)

Do one section of the room at a time. The tiles are simply maneuvered into position and pressed into place. See Fig. 3-48.

Border tiles for the edges of the room can be easily cut to size with a pair of ordinary household shears. See Fig. 3-49.

It doesn't take long for the room to take shape. There is no smell from the adhesive. The floor can be used as soon as it is finished. This type of floor replacement or repair is commonplace today.

Paneling a Room

The room to be paneled may have cracked walls. This means you'll need furring strips to cover the old walls and provide good nailing and shimming for a smooth wall. See Fig. 3-50. Apply furring strips (1×1s or 1×3s) vertically at 16-inch intervals for full-size 4-foot $\times$ 8-foot panels. Apply the furring strips horizontally at 16-inch intervals for random-width paneling. Start at the end of the wall farthest from the main entrance to the room. The first panel should be plumbed from the corner by striking a line 48 inches out. Trim the panel in the corner (a rasp works well) so that the panel aligns on the plumb line. Turn the corner, and butt the next panel to the first panel. Plumb in the same manner as the first panel.

To make holes in the paneling for switch boxes and outlets, trace the box's outline in the desired place on the panel. Then drill holes at the four corners of the area marked. Next, cut between the holes with a key-hole saw. Then rasp the edges smooth. See Fig. 3-51.

Before you apply the adhesive, make sure the panel is going to fit. If the panel is going to be stuck right onto the existing wall, apply the adhesive at 16-inch intervals, horizontally and vertically. See Fig. 3-52A.

Spacing   Avoid a tight fit. Above grade, leave a space approximately the thickness of a matchbook cover at the sides and a ⅛-inch space at top and bottom. Below grade, allow ¼ inch top and bottom. Allow not less than a ¹⁄₁₆-inch space (the thickness of a dime) between panels in high-humidity areas. See Fig 3-52B.

Vapor barrier   A vapor barrier is needed if the panels are installed over masonry walls. It doesn't matter if the wall is above or below grade. See Fig. 3-52C. If

Fig. 3-48 *Do one section of the room at a time. The tiles can be maneuvered into position and pressed into place.* (Armstrong Cork)

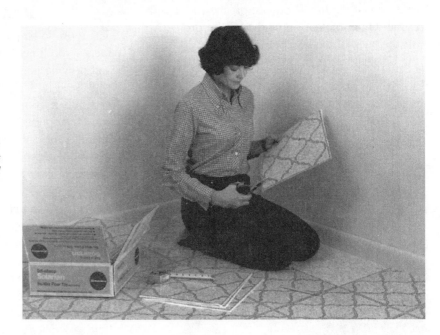

Fig. 3-49 *Border tiles for the edges of the room can be easily cut to size with a pair of household shears.* (Armstrong Cork)

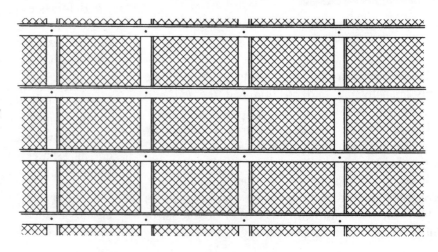

Fig. 3-50 *Applying vertical and horizontal furring strips on an existing wall.* (Valu)

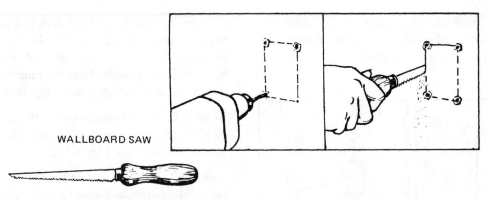

WALLBOARD SAW

Fig. 3-51 *Cutting holes in the paneling for electrical switches and outlets.* (Valu)

(A)

1/8″ OR 1/4″

MATCHBOOK COVER THICKNESS

1/8″ OR 1/4″

(B)

(C)

Fig. 3-52 *(A) Make sure the panel fits before you apply adhesive* (Valu)*. (B) Allow space at top and bottom and at each side for the panel to expand* (Valu)*. (C) Apply a vapor barrier to the walls before you place the furring strips onto the wall.* (Abitibi)

the plaster is wet or the masonry is new, wait until it is thoroughly dry. Then condition the panels to the room.

Adhesives In Fig. 3-53 a caulking gun is being used to apply the adhesive to the furring strips. In some cases it is best to apply the adhesive to the back of the panel. Follow the directions on the tube. Each manufacturer has different instructions.

Hammer a small nail through the top of the panel. It should act as a hinge. Press the panel against the wall or frame to apply glue to both sides. See Fig. 3-54A.

Pull out the bottom of the panel and keep it about 8 to 10 inches away from the wall or framework. Use a

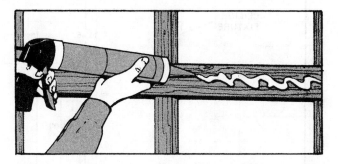

Fig. 3-53 *Use a caulking gun to apply the adhesive.* (Valu)

wood block. See Fig. 3-54B. Let the adhesive become tacky. Leave it for about 8 to 10 minutes.

CAUTION: Methods of application for some types of panel adhesives may differ. Always check the instructions on the tube.

Fig. 3-54 *(A) Hammer a nail through the top of the panel. This will act as a hinge so that the panel can be pressed against the wall to apply the glue to both surfaces* (Valu). *(B) Pull out the bottom of the panel and keep it about 8 to 10 inches away from the wall with a wood block until the adhesive becomes tacky* (Valu). *(C) Use a compass to mark panels so that they fit perfectly against irregular surfaces* (Abitibi). *(D) Use a hammer and block of wood covered with a cloth to spread the adhesive evenly.* (Valu)

Let the panel fall back into position. Make sure that it is correctly aligned. The first panel should butt one edge against the adjacent wall. The panel should be completely plumb. Trim the inner edge as needed. This is done so that the outer edge falls on a stud.

Scribing Use a compass to mark panels so that they fit perfectly. It can be used to mark the variations in a butting surface. See Fig. 3-54C.

Nailing Use finishing nails (3d) or brads (1¼-inch) or annular hardboard nails (1-inch). Begin at the edge. Work toward the opposite side. Never nail opposite ends first, then the middle. With a hammer and a cloth-covered block, hammer gently to spread the adhesive evenly. See Fig. 3-54D. If you are using adhesive alone to hold the panels in place, don't remove the nails until the adhesive is thoroughly dry. Then, after they're out, fill the nail holes with a matching putty stick.

Moldings can be glued into place or nailed.

Installing a Ceiling

There are a number of methods used to install a new ceiling in a basement or a recreation room. In fact, some very interesting tiles are available for living rooms as well.

Replacing an old ceiling. The first thing to do in replacing an old, cracked ceiling is to lay out the room. See Fig. 3-55. It should be laid out accurately to scale. Use ½" = 1'0" as a scale. Do the layout on graph paper.

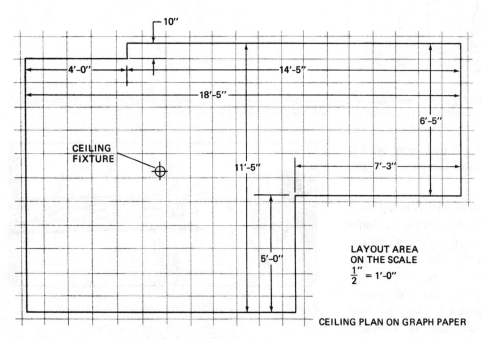

Fig. 3-55 *Lay out your room accurately to scale on graph paper.* (Grossman Lumber)

Then, on tracing paper, draw ¹/₂-inch squares representing the 12-inch ceiling tile. Lay the tracing paper over the ceiling plan. Adjust the paper until the borders are even. Border pieces should never be less than half a tile wide. Use cove molding at the walls to cover the trimmed edges of the tile. See Fig. 3-56. Tile can be stapled or applied with mastic. This is especially true of fiber tile. It has flanges that will hold staples. See Fig. 3-57. You can staple into wallboard or into furring strips nailed to joists. When you apply tile to furring strips, follow your original layout but start in one corner and work toward the opposite corner.

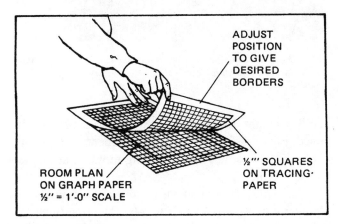

Fig. 3-56 *Draw ¹/₂-inch squares on tracing paper. Lay the tracing paper over the ceiling plan. Adjust the paper until the borders are even. Border pieces should never be less than half a tile wide.* (Grossman Lumber)

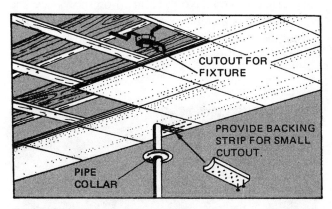

Fig. 3-57 *Fiber tile is stapled to furring strips applied to the joists.* (Grossman Lumber)

If you apply tile with mastic, snap a chalk line, as shown in Fig. 3-58. This will find the exact center of your ceiling. In applying the mastic to the ceiling tile, put a golf-ball-sized blob of mastic on each corner of the tile and one in the middle. See Fig. 3-59.

Apply the first tile where the lines cross. Then work toward the edges. Set the tile just out of position, then

Fig. 3-58 *If you are going to glue the tiles to the ceiling, snap a chalk line to find the exact center of the ceiling.* (Grossman Lumber)

Fig. 3-59 *Place a blob of mastic at each corner and in the middle of the tile before you slide it into place on the ceiling surface.* (Grossman Lumber)

slide it into place, pressing firmly to ensure a good bond and a level ceiling. See Fig. 3-60. Precut holes for lighting fixtures and pipes. Use a sharp utility knife. Always be sure to make the cutouts with the tile face up. See Fig. 3-61.

The drop ceiling   If an old ceiling is too far damaged to be repaired inexpensively, it might be best to install a drop ceiling. This means the ceiling will be completely new and will not rely upon the old ceiling for support. This way the old ceiling does not have to be repaired.

Figure 3-62 shows the first operation of a drop ceiling. Nail the moldings to all four walls at the desired

Fig. 3-60 *Apply the first tile where the lines cross on the ceiling. Work toward the edges.* (Grossman Lumber)

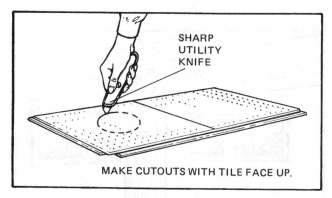

SHARP UTILITY KNIFE

MAKE CUTOUTS WITH TILE FACE UP.

Fig. 3-61 *Precut holes for lighting fixtures and pipes. Use a sharp utility knife. Always be sure to cut the tiles with the face up.* (Grossman Lumber)

Fig. 3-62 *For a new suspended ceiling, nail molding to all four walls.* (Armstrong Cook)

Fig. 3-63 *A suspended ceiling in a basement is begun the same way. Nail molding on all four walls.* (Armstrong Cook)

ceiling height. Either metal or wood moldings may be used. Figure 3-63 shows how a basement drop ceiling is begun the same way, by nailing molding at the desired height.

The next step is to install the main runners on hanger wires as shown in Fig. 3-64. The first runner is always located 26 inches out from the sidewall. The remaining units are placed 48 inches O.C., perpendicular to the direction of the joists. Unlike conventional suspended ceilings, this type requires no complicated measuring or room layout. Figure 3-65 shows how the main runner is installed by fastening hanger wires at 4-foot intervals—the conventional method for a basement drop ceiling.

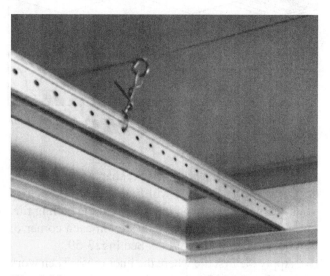

Fig. 3-64 *Main runners are installed on hanger wires.* (Armstrong Cook)

Fig. 3-65 *In the basement installation, install the main runners by fastening them to hanger wires at 4-foot intervals.* (Armstrong Cook)

After all main runners are in place, begin installing ceiling tile in a corner of the room. Simply lay the first 4 feet of tile on the molding, snap a 4-foot cross T onto the main runner, and slide the T into a special concealed slot on the leading edge of the tile. See Fig. 3-66. This will provide a ceiling without the metal supports showing.

Fig. 3-66 *After all the main runners are in place, begin installing ceiling tile in a corner of the room. Note how the T slides into the tile and disappears.* (Armstrong Cook)

In the conventional method, as shown in the basement installation in Fig. 3-67, cross T's are installed between main runners. The T's have tabs that engage slots in the main runner to lock firmly in place. After this is done, the tiles are slid into place from above the metal framework and allowed to drop into the squares provided for them.

Fig. 3-67 *When the main runners are in place, install cross T's between them. The T's have tabs that engage slots in the main runner to lock firmly in place. This type of suspended ceiling will have the metal showing when the job is finished.* (Armstrong Cook)

In Fig. 3-68 tile setting is continuing. The tiles and cross T's are inserted as needed. Note how all metal suspension members are hidden from view in the finished portion of the ceiling.

The result of this type of ceiling is an uninterrupted surface. There is no beveled edge to produce a line across the ceiling. All supporting ceiling metal is concealed.

One of the advantages of dropping a ceiling is the conservation of energy. Heat rises to the ceiling. If the ceiling is high, the heat is lost to the room. The lower the ceiling, the less the volume to be heated and the less energy needed to heat the room.

Fig. 3-68 *Continue across the room in this manner. Insert the tiles and cross T's.* (Armstrong Cook)

Replacing an Outside Basement Door

Many houses have basements that can be made into playrooms or activity areas. The workshop that is needed but can't be put anywhere else will probably wind up in the basement. Basements can be very difficult to get out of if a fire starts around the furnace or hot water heater area. One of the safety measures that can be taken is to place a doorway directly to the outside so you don't have to try to get up a flight of stairs and then through the house to escape a fire.

Figure 3-69 shows how an outside stairway can be helpful for any basement. This type of entrance or exit from the basement can be installed in any house. It takes some work, but it can be done.

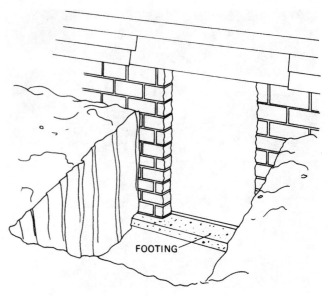

Fig. 3-70 *Excavate the area around the proposed entrance.* (Bilco)

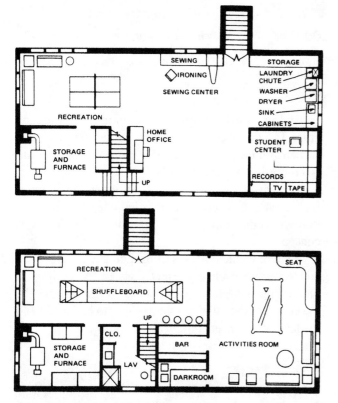

Fig. 3-69 *Note the location of the outside basement door or entry on these drawings.* (Bilco)

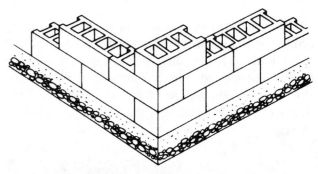

Fig. 3-71 *Start the blocks on top of the footings.* (Bilco)

Figure 3-70 shows how digging a hole and breaking through the foundation in stages will permit using tools to get the job done. A basement wall of concrete blocks is easier to break through than a poured wall.

Start by building the areaway. This is done by laying a 12- to 16-inch concrete footing. This footing forms a level base for the first course of concrete block. See Fig. 3-71. Allow the footing to set for 2 to 3 days before you lay blocks for the walls.

The areaway for a size C door is shown in Fig. 3-72. Figure 3-72A shows the starting course. Figure 3-72B shows how the next course of blocks is laid. The top course should come slightly above ground level. See Fig. 3-73. It should be about 3 inches from the required areaway height as given in the construction guide provided with the door.

Build up the areaway to the right height in the manner shown. You may want to waterproof the outside of the new foundation. Use the material recommended by your local lumber dealer or mason yard.

Now build a form for capping the wall. See Fig. 3-74. Fill the cores of the top blocks halfway up with crushed balls of newspaper or insulation. The cap to be poured on this course will bring the areaway up to the required height.

When the cap is an inch or so below the desired height, set the door back in position with the header flange between the siding and the sheathing underneath. Make sure the frame is square. Insert the mounting screws with the spring-steel nuts in the side pieces and the sill. Embed them in the wet concrete to hold the

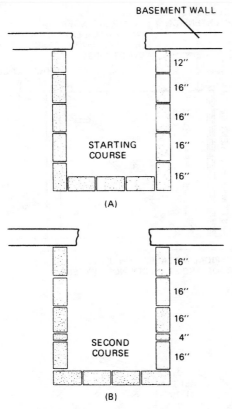

BASEMENT WALL

12"
16"
16"
16"
16"

STARTING
COURSE

(A)

16"
16"
16"
4"
16"

SECOND
COURSE

(B)

Fig. 3-72 *(A) The starting course, looking down into the hole.* (Bilco) *The second course, on top of the first, looks like this.* (Bilco)

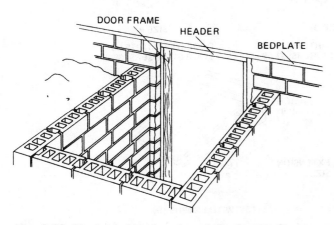

DOOR FRAME
HEADER
BEDPLATE

Fig. 3-73 *The finished block work ready for the cap of concrete.* (Bilco)

screws tightly. See Fig. 3-75. Continue pouring the cap. Bring the concrete flush with the bottom of the sill and side-piece flanges. Do not bring the capping below the bottom of the door. The door should rest on top of the foundation. With a little extra work, the cap outside the door can be chamfered downward as shown in Fig. 3-75. This ensures good drainage. Trowel the concrete smooth and level.

Install a prehung door in the wall of the basement. These come in standard sizes. The door should be

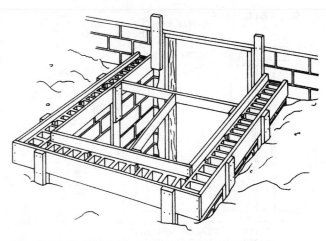

Fig. 3-74 *Forms for the concrete cap are built around the concrete block.* (Bilco)

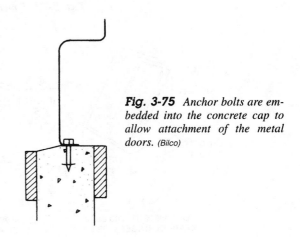

Fig. 3-75 *Anchor bolts are embedded into the concrete cap to allow attachment of the metal doors.* (Bilco)

selected to fit the hole made in the wall. Use the widest standard unit that fits the entryway you have built.

Figure 3-76 shows how the steps are installed in the opening. The stringers for the steps are attached to the walls of the opening. See Fig. 3-77.

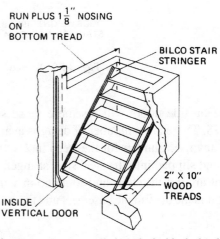

RUN PLUS 1⅛" NOSING
ON
BOTTOM TREAD

BILCO STAIR
STRINGER

2" X 10"
WOOD
TREADS

INSIDE
VERTICAL DOOR

Fig. 3-76 *The stairs are installed inside the blocked-in hole.* (Bilco)

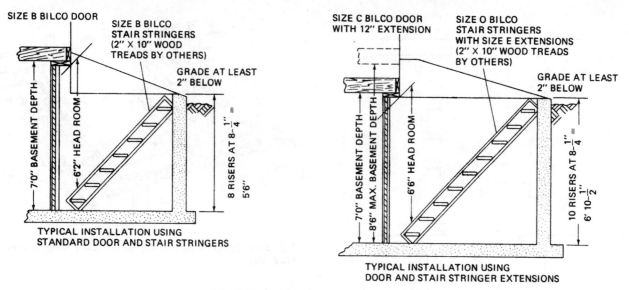

Fig. 3-77 *Typical door installations.* (Bilco)

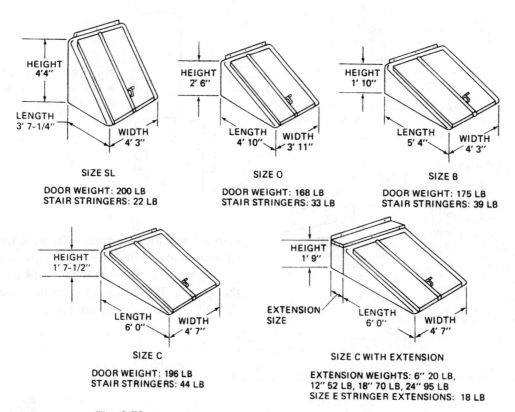

SIZE SL
DOOR WEIGHT: 200 LB
STAIR STRINGERS: 22 LB

SIZE O
DOOR WEIGHT: 168 LB
STAIR STRINGERS: 33 LB

SIZE B
DOOR WEIGHT: 175 LB
STAIR STRINGERS: 39 LB

SIZE C
DOOR WEIGHT: 196 LB
STAIR STRINGERS: 44 LB

SIZE C WITH EXTENSION
EXTENSION WEIGHTS: 6" 20 LB,
12" 52 LB, 18" 70 LB, 24" 95 LB
SIZE E STRINGER EXTENSIONS: 18 LB

Fig. 3-78 *Various shapes and sizes of outside basement doors.* (Bilco)

The outside doors will resemble those shown in Fig. 3-78. There are a number of designs available for almost any use. Lumber for the steps in 2 × 10s cut to length and slipped into the steel stair stringers.

Seal around the door and foundation with caulk. Seal around the door in the basement and the wall. Allow the stairwell to *air out* during good weather by keeping the outside doors open. This will allow the moisture from the masonry to escape. After it has dried out, the whole unit will be dry.

4
CHAPTER

Converting and Adding Space

CONVERTING EXISTING SPACES

There is never enough room in any house. All it takes is a few days after you unpack all your belongings to find out that there isn't enough room for everything. The next thing new homeowners do is look around at the existing building to see what space can be converted to other uses. It is usually too expensive to add on immediately, but it is possible to remodel the kitchen, porch, or garage.

Adding a Bathroom

As the family grows, the need for more bathrooms becomes very apparent. First you look around for a place to put the bathroom. Then you locate the plumbing and check to see what kind of a job it will be to hook up the new bathroom to the cold and hot water and to the drains. How much effort will be needed to hook up pipes and run drains? How much electrical work will be needed?

These questions will have to be answered as you plan for the additional bathroom.

Start by getting the room measurements. Then make a plan of where you would place the various necessary items. Figure 4-1 shows some possibilities. Your plan can be as simple as a lavatory and water closet, or you can expand with a shower, a whirlpool bath, and a sauna. The amount of money available will usually determine the choice of fixtures.

Look around at books and magazines as well as the literature of manufacturers of bathroom fixtures. Get some ideas as to how you would rearrange your own bathroom or make a new one.

If you have a larger room to remodel and turn into a bathroom, you might consider what was done in Fig. 4-2. This is a Japanese bathroom dedicated to the art of bathing. Note how the tub is fitted with a shower to allow you to soap and rinse on the bathing platform before soaking, as the Japanese do.

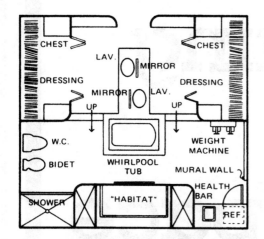

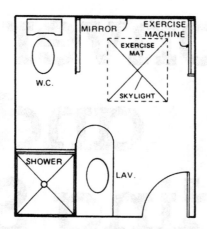

Fig. 4-1 *Floor plans for bathrooms.* (Kohler)

Fig. 4-2 *Japanese bathroom design. The platform gives the illusion of a sunken bath. Many of these features can be incorporated into a remodeled room which can serve as a bathroom.* (Kohler)

You may not want to become too elaborate with your new bathroom. All you have to do then is decide how and what you need. Then draw your ideas for arrangements. Check the plumbing to see if your idea will be feasible. Order the materials and fixtures and then get started.

Providing Additional Storage

Cedar-lined closet   Aromatic red cedar closet lining is packed in a convenient, no-waste package which contains 20 square feet. It will cover 16⅓ square feet of

Fig. 4-3 *Cedar-lined closet. The aromatic red cedar serves as a moth deterrent.* (Grossman Lumber)

wall space for lifetime protection from moths. To install the cedar, follow these steps. See Fig. 4-3.

1. Measure the wall, ceiling, floor, and door area of the closet. Figure the square footage. Use the length times width to produce square feet.

2. Cedar closet lining may be applied to the wall either vertically or horizontally. See Fig. 4-4. When applied to the rough studs, pieces of cedar must be applied horizontally.

3. When you apply the lining to the wall, place the first piece flush against the floor with grooved edge down. Use small finishing nails to apply the lining to the wall. See Fig. 4-5.

Fig. 4-4 *Placing a piece of cut-to-fit red cedar lining vertically in a closet.* (Grossman Lumber)

Fig. 4-5 *Work from the bottom up when you apply the cedar boards.* (Grossman Lumber)

4. When you are cutting a piece of cedar to finish out a course or row of boards, always saw off the tongued end so that the square sawed-off end will fit snugly into the opposite corner to start the new course. See Fig. 4-6.

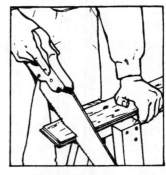

Fig. 4-6 *To start a new course, saw off the tongued end. The square sawed end will fit snugly into the opposite corner.* (Grossman Lumber)

5. Finish one wall before you start another. Line the ceiling and floor in a similar manner. Each piece of cedar is tongue-and-grooved for easy fit and application. See Fig. 4-7.

The cedar-lined closet will protect your woolens for years. However, in some instances you might not have a closet to line. You might need extra storage space. In this instance take a look at the next section.

Fig. 4-7 *Finish one wall before you start another.* (Grossman Lumber)

Building extra storage space Any empty corner can provide the back and one side of a storage unit. This simple design requires a minimum of materials for a maximum of storage. Start with a 4-foot unit now and add 2 feet later. See Fig. 4-8.

Use a carpenter's level and square to check a room's corners for any misalignment. Note where the irregularities occur. The basic frame is built from 1-inch and 2-inch stock lumber. It can be adjusted to fit any irregularities in the walls or floor. Nail the top and shelf cleats into the studs. Cut the shelf from $\frac{1}{2}$-inch plywood. Slip it into place and nail it to the cleat. Attach the clothes rod with the conventional brackets. Fit, glue, and nail the floor, side, and top panels. Make them from $\frac{1}{8}$-inch hardboard. You can buy metal or hardwood sliding doors

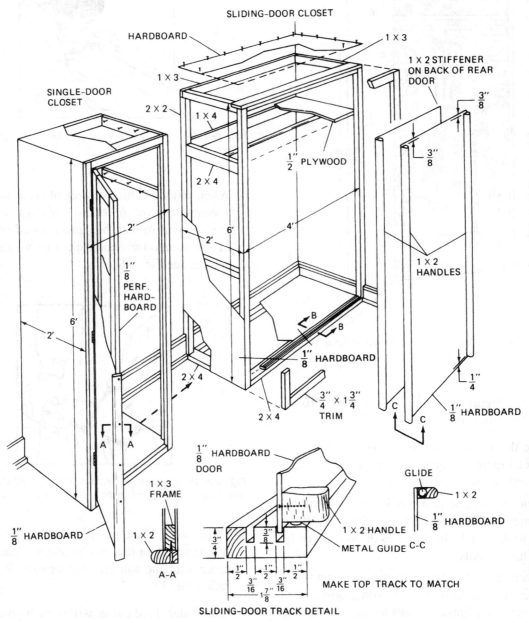

Fig. 4-8 *Extra storage space. The sliding-door closet and the single-door closets.* (Grossman Lumber)

and tracks. Or you can make the doors and purchase only the tracks. Doors are made of single sheets of ⅛-inch hardboard stiffened with full-length handles of 1 × 2 trim set on the edge. Metal glides on the ends of handles carry the weight of the door and prevent binding.

Figure 4-9 shows some of the details for making the door operational.

A swinging-door unit is built almost like the double-door unit. The framework should be fastened to the wall studs where possible. Make the door frame of 1 × 3s laid flat with ⅛-inch hardboard (plain) on the face and ⅛-inch perforated hardboard on the inside. Again, a full-length 1 × 2 makes the handle.

Attach three hinges to the door. Make sure that the pins line up so that the door swings properly. Place the door in the opening and raise it slightly. Mark the frame and chisel out notches for the hinges.

Other types of storage space For adequate, well-arranged storage space, plan your closets first. Minimum depth of closets should be 24 inches. The width can vary to suit your needs. But provide closets with large doors for adequate access. See Fig. 4-10.

Figure 4-11 shows some of the arrangements for closets. Use 2 × 4 framing for dividers and walls so that shelves may be attached later. Carefully measure and fit ⅜-inch gypsum board panels in place. Finish the interior or outside of the closet as you would any type of drywall installation. Install the doors you planned for. These can be folding, bifold, or a regular prehung type.

Remodeling a Kitchen

In remodeling a kitchen the major problem is the kitchen cabinets. There are any number of these available ready made. They can be purchased and installed to make any type of kitchen arrangement desired.

The countertops have already been covered. The color of the countertop laminate must be chosen so that it will blend with the flooring and the walls. This is the job of the person who will spend a great deal of time in the kitchen.

Planning the kitchen The kitchen begins with a set of new cabinets for storage and work areas. The manufacturer of the cabinets will supply complete installation instructions.

Check the drawing and mark on the walls where each unit is to be installed. Mark the center of the stud lines. Mark the top and the bottom of the cabinet so that you can locate them easily at the time of installation.

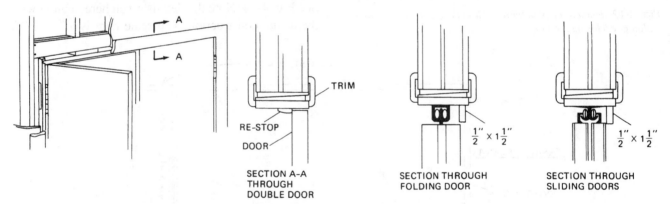

Fig. 4-9 *Details of the door installation for the closets in Fig. 4-8.* (Grossman Lumber)

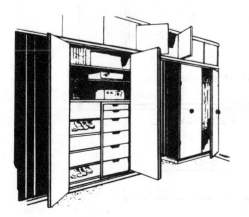

Fig. 4-10 *A variation of cabinet storage designs.* (Grossman Lumber)

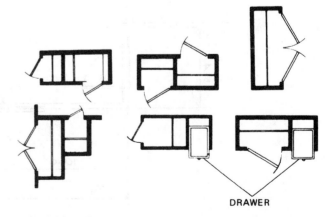

Fig. 4-11 *Different plans and door openings for closets.* (Grossman Lumber)

As soon as the installation is completed, wipe the cabinets with a soft cloth dampened with water. Dry the cabinets immediately with another clean soft cloth. Follow this cleaning with a very light coat of high-quality liquid or paste wax. The wax helps keep out moisture and causes the cabinets to wear longer.

Finishing up the kitchen   After the cabinets have been installed, it is time to do the plumbing. Have the sink installed and choose the proper faucet for the sink.

Kitchen floor   Now it's time to put down the kitchen floor. In most cases the flooring preference today is carpet, although linoleum and tile are also used. It is easier to clean carpet—just vacuuming is sufficient. It is quieter and can be wiped clean easily if something spills. If a total remodeling job has been done, it may be a good idea to paint or wallpaper the walls before installing the flooring. A new range and oven are usually in order, too. The exhaust hood should be properly installed electrically and physically for exhausting cooking orders and steam. This should complete the kitchen remodeling. Other accents and touches here and there are left up to the user of the kitchen.

Enclosing a Porch

One of the first things to do is to establish the actual size you want the finished porch to be. In the example shown here, a patio (16 feet × 20 feet) is being enclosed. A quick sketch will show some of the possibilities (Fig. 4-12). This becomes the foundation plan. It is drawn ¼ inch to equal 1 foot. The plan can then be used to obtain a building permit from the local authorities.

The floor plan is next. See Fig. 4-13. It shows the location of the doors and windows and specifies their sizes.

A cross section of the addition or enclosure is next. See Fig. 4-14. Note the details given here dealing with the actual construction. The scale here is ½″ = 1′0″.

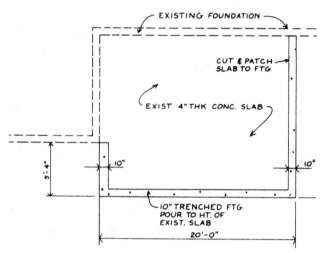

Fig. 4-12   *Foundation plan for an addition to an already existing building; enclosing a patio.*

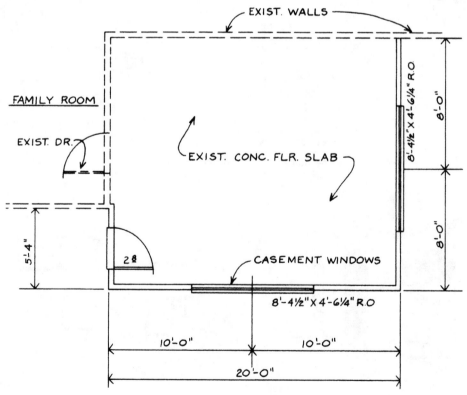

Fig. 4-13   *Floor plan for an addition.*

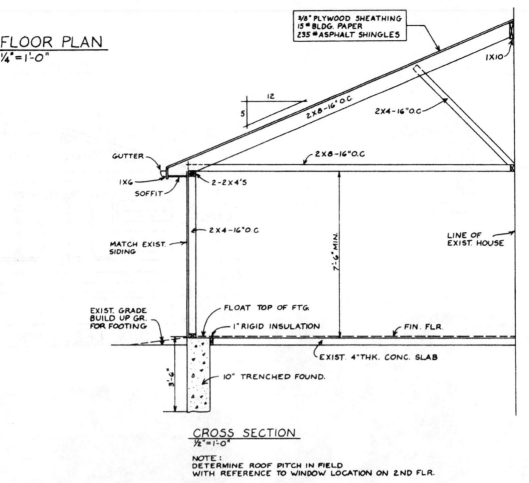

FLOOR PLAN
1/4"=1'-0"

3/8" PLYWOOD SHEATHING
15 # BLDG. PAPER
235 # ASPHALT SHINGLES

1X10

12

5

2X8-16" O.C

2X4-16"O.C.

2X8-16"O.C

GUTTER

1X6

SOFFIT

2-2X4'S

MATCH EXIST.
SIDING

2X4-16"O.C.

7'-6" MIN.

LINE OF
EXIST. HOUSE

EXIST. GRADE
BUILD UP GR.
FOR FOOTING

FLOAT TOP OF FTG.

1" RIGID INSULATION

FIN. FLR.

EXIST. 4"THK. CONC. SLAB

3'-6"

10" TRENCHED FOUND.

CROSS SECTION
1/2"=1'-0"

NOTE:
DETERMINE ROOF PITCH IN FIELD
WITH REFERENCE TO WINDOW LOCATION ON 2ND FLR.

Fig. 4-14 *Cross section view of the addition.*

Note that the roof pitch is to be determined with reference to the window location on the second floor of the existing building. Figure 4-17A shows what actually happened to the pitch as determined by the window on the second floor. As you can see from the picture, the pitch is not 5:12 as called for on the drawing. Because of the long run of 16 feet, the 2 × 6 ceiling joists, 16 O.C., had to be changed to 2 × 10s. This was required by the local code. It was a good requirement, since with the low slope on the completed roof, the pile-up of snow would have caused the roof to cave in. See Fig. 4-17A.

Elevations Once the floor plan and the foundation plan have been completed, you can begin to think about how the porch will look enclosed. This is where the elevation plans come in handy. They show you what the building will look like when it is finished. Figure 4-15A shows how part of the porch will look when it is extended past the existing house in the side elevation.

Figure 4-15B shows how the side elevation looks when finished. Note the storm door and the outside light for the steps.

The rear elevation is simple. It shows the five windows that allow a breeze through the porch on days when the windows can be opened. See Fig. 4-16 for a view of the enclosed porch viewed from the rear.

A side elevation is necessary to see how the other side of the enclosure will look. This shows the location of the five windows needed on this side to provide ventilation. Figure 4-17A illustrates the way the enclosure should look. Figure 4-17B shows how the finished product looks with landscaping and the actual roofline created by the second-floor window location. With this low-slope roof, heating cables must be installed on the roof overhang. This keeps ice jams from forming and causing leaks inside the enclosed porch.

Once your plans are ready and you have all the details worked out, it is time to get a building permit. You have to apply and wait for the local board's decision. If you comply with all the building codes, you can go ahead. This can become involved in some communities. The building permit shown in Fig. 4-18 shows some of the details and some of the people involved in

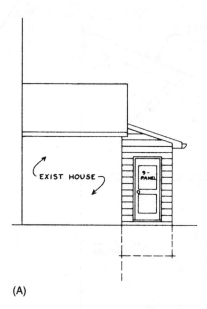

(A)

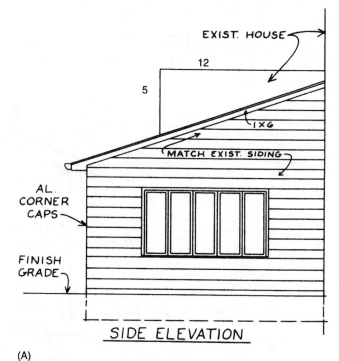

SIDE ELEVATION

(A)

(B)

Fig. 4-15 *(A) Side elevation of the addition. (B) What the side will look like when finished.*

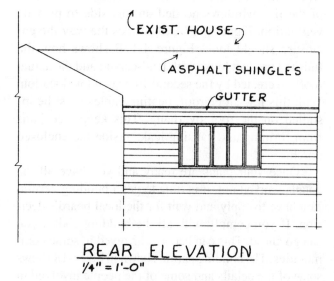

REAR ELEVATION

1/4" = 1'-0"

Fig. 4-16 *Rear elevation of the addition.*

(B)

Fig. 4-17 *(A) Side elevation of the addition. (B) What the side elevation will look like when finished.*

issuing a building permit. This building permit is for the porch enclosure shown in the previous series of drawings and pictures.

Starting the project Once you have the building permit, you can get started. The first step is to dig the hole for the concrete footings or for the trench-poured foundation. Once the foundation concrete has set up, you can add concrete blocks to bring the foundation up to the existing level. After the existing grade has been established with the addition, you can proceed as usual for any type of building. Put down the insulation strip and the soleplate. Attach the supports to the existing wall and put in the flooring

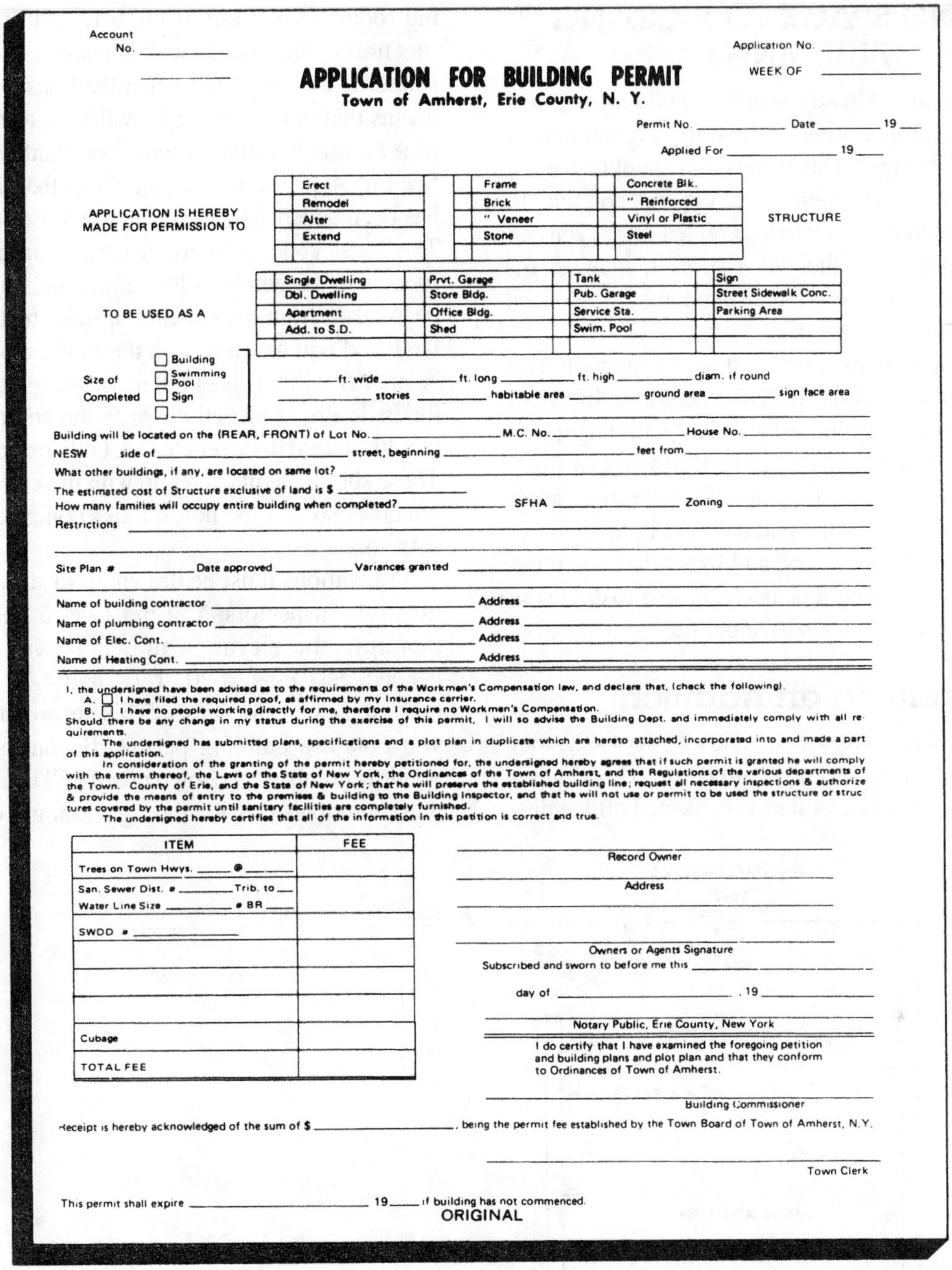

APPLICATION FOR BUILDING PERMIT
Town of Amherst, Erie County, N. Y.

Account No. _____

Application No. _____

WEEK OF _____

Permit No. _____ Date _____ 19 ___

Applied For _____ 19 ___

APPLICATION IS HEREBY MADE FOR PERMISSION TO

			STRUCTURE
☐ Erect	☐ Frame	☐ Concrete Blk.	
☐ Remodel	☐ Brick	☐ " Reinforced	
☐ Alter	☐ " Veneer	☐ Vinyl or Plastic	
☐ Extend	☐ Stone	☐ Steel	

TO BE USED AS A

☐ Single Dwelling	☐ Prvt. Garage	☐ Tank	☐ Sign
☐ Dbl. Dwelling	☐ Store Bldg.	☐ Pub. Garage	☐ Street Sidewalk Conc.
☐ Apartment	☐ Office Bldg.	☐ Service Sta.	☐ Parking Area
☐ Add. to S.D.	☐ Shed	☐ Swim. Pool	

Size of Completed ☐ Building ☐ Swimming Pool ☐ Sign ☐

_____ ft. wide _____ ft. long _____ ft. high _____ diam. if round

_____ stories _____ habitable area _____ ground area _____ sign face area

Building will be located on the (REAR, FRONT) of Lot No. _____ M.C. No. _____ House No. _____

NESW side of _____ street, beginning _____ feet from _____

What other buildings, if any, are located on same lot? _____

The estimated cost of Structure exclusive of land is $ _____

How many families will occupy entire building when completed? _____ SFHA _____ Zoning _____

Restrictions _____

Site Plan # _____ Date approved _____ Variances granted _____

Name of building contractor _____ Address _____

Name of plumbing contractor _____ Address _____

Name of Elec. Cont. _____ Address _____

Name of Heating Cont. _____ Address _____

I, the undersigned have been advised as to the requirements of the Workmen's Compensation law, and declare that, (check the following)
A. ☐ I have filed the required proof, as affirmed by my Insurance carrier.
B. ☐ I have no people working directly for me, therefore I require no Workmen's Compensation.
Should there be any change in my status during the exercise of this permit, I will so advise the Building Dept. and immediately comply with all requirements.
The undersigned has submitted plans, specifications and a plot plan in duplicate which are hereto attached, incorporated into and made a part of this application.
In consideration of the granting of the permit hereby petitioned for, the undersigned hereby agrees that if such permit is granted he will comply with the terms thereof, the Laws of the State of New York, the Ordinances of the Town of Amherst, and the Regulations of the various departments of the Town, County of Erie, and the State of New York; that he will preserve the established building line; request all necessary inspections & authorize & provide the means of entry to the premises & building to the Building Inspector, and that he will not use or permit to be used the structure or structures covered by the permit until sanitary facilities are completely furnished.
The undersigned hereby certifies that all of the information in this petition is correct and true.

ITEM	FEE
Trees on Town Hwys. _____ @ _____	
San. Sewer Dist. # _____ Trib. to _____	
Water Line Size _____ # BR	
SWDD # _____	
Cubage	
TOTAL FEE	

Record Owner _____

Address _____

Owners or Agents Signature _____

Subscribed and sworn to before me this _____

day of _____ , 19 _____

Notary Public, Erie County, New York

I do certify that I have examined the foregoing petition and building plans and plot plan and that they conform to Ordinances of Town of Amherst.

_____ Building Commissioner

Receipt is hereby acknowledged of the sum of $ _____ , being the permit fee established by the Town Board of Town of Amherst, N.Y.

_____ Town Clerk

This permit shall expire _____ 19 ___ if building has not commenced.

ORIGINAL

Fig. 4-18 *Application for a building permit.*

joists. Once the flooring is in, you can build the wall framing on it and push the framing up and into place. Put on the rafters and the roof. Get the whole structure enclosed with a nail base or plywood on the outside walls. Place the windows and doors in. Put on the roof and then the siding. Trim the outside around the door and windows. Check the overhang and place the trim where it belongs. Put in the gutters and connect them to the existing system. If there is to be electrical wiring, put that in next. In most porches there is no plumbing, so you can go on to the drywall. Finish the ceiling and walls.

After the interior is properly trimmed, it should be painted. The carpeting or tile can be placed on the floor and the electrical outlets covered with the proper plates. The room is ready for use.

Finishing the project Throughout the building process, the local building inspector will be making calls to see that the code is followed. This is for the benefit of both the builder and the owner.

There is another *benefit* (to the local government) of the inspections and the building permit. Once the addition is inhabited, the structure can go on the tax roles and the property tax can be adjusted accordingly.

ADDING SPACE TO EXISTING BUILDINGS

Adding space to a already standing building requires some special considerations. You want the addition to look as if it "belongs." This means you should have the siding the same as the original or as close as possible. In some cases you may want it to look added-on, so you can contrast the new with the old. However, in most instances, the intention is to match up the addition as closely as possible.

First, the additional space must have a function. You may need it for a den, an office, or a bedroom. In any of these instances you don't need water or plumbing if you already have sufficient bathroom facilities on that floor. You will need electrical facilities. Plan the maximum possible use of the building and then put in the number of electrical outlets, switches, and lights that you think you can use. Remember, it is much cheaper to do it now than after the wallboard has been put up.

Planning an Addition

In the example used here, we will add a 15-foot × 22-foot room onto an existing, recently built house. The addition is to be used as an office or den. It is located off the din-ing room, so it is out of the way of through-the-house foot traffic. The outside wall will also serve to deaden the sound. It was insulated when the house was built. This means that only three walls will have to be constructed. Make a rough sketch of what you think will be needed. See Fig. 4-19 for an example. Note that the light switch has been added. The windows are of two different sizes. The local code calls for a certain amount of square footage in the windows for ventilation. One window has been taken from the existing upstairs bedroom. It is too large and will interfere with the roofline of this addition. A smaller window is put in up there, and the one used in the bedroom is moved down to the addition. Only two windows need to be purchased. Only one door is needed. These should be matched up with the existing doors and windows so that the house looks complete from the inside, too.

Elevations must be drawn up to show to the building code inspectors. You must get a building permit; therefore the elevation drawings will definitely be needed. See Figs. 4-20, 4-21, and 4-22 for the front, rear, and side elevations. This information will help you obtain a building permit. If you use Andersen's window numbers, the town board will be able to see that these windows provide the right amount of ventilation.

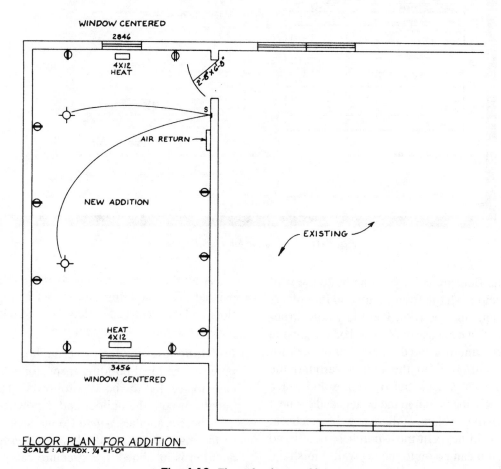

Fig. 4-19 *Floor plan for an addition.*

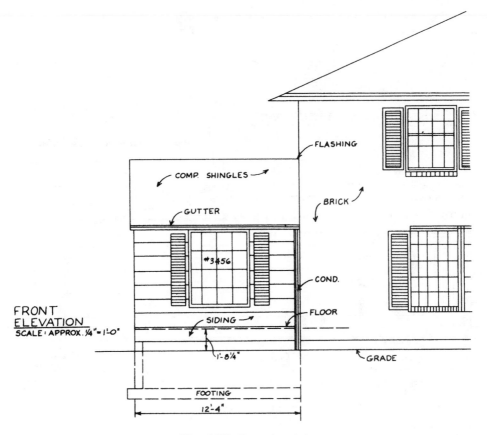

Fig. 4-20 *Front elevation.*

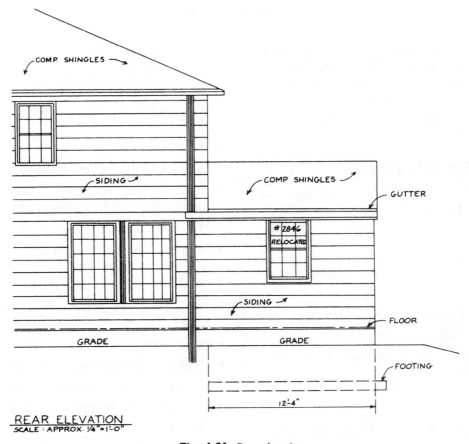

Fig. 4-21 *Rear elevation.*

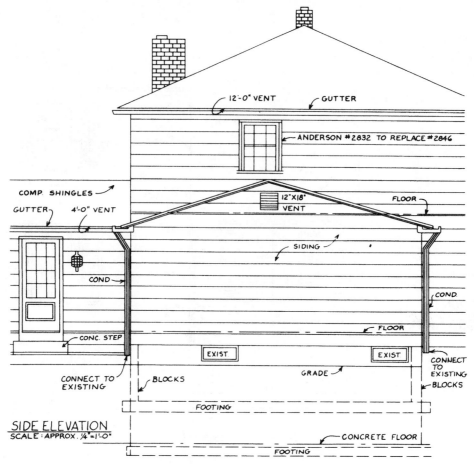

Fig. 4-22 *Side elevation.*

To make the finished product fit your idea of what you wanted, it is best to write a list of specifications. Make sure you list everything you want done and how you want it done. For example, the basement, siding, overhang, flooring, windows, door, walls, roof, and even the rafters should be specified.

Specifications

Addition for 125 Briarhill in Town of Amherst. (See detailed sketches for rear, front, and side elevations.)

Basement

- Crawl space—skim coat of concrete over gravel. Drainage around footings and blocks to prevent moisture buildup.
- Blocks on concrete footings—42 inches deep.
- To be level with adjoining structure.
- To be waterproofed. Fill to be returned and leveled around the exterior.

Siding

- National Gypsum Woodrock Prefinished (as per existing).

- Size to match existing.
- Must match at corners and overlap in rear to look as if built with original structure.
- To be caulked around windows.

Overhang

- To match existing as per family room extension.
- Gutters front and rear to match existing. Downspouts (conduits) to match existing and to connect.
- Ventilation screen in Upson board overhang.
- Exterior fascia and molding to match existing.

Flooring

- Kiln-dried 2 × 10 (construction or better) on 16-inch centers.
- Plywood subfloor, ⅝-inch exterior, A–D.
- Plywood subfloor, ½-inch A–D interior.
- Fiberglas insulation strip between soleplate and blocks.
- Must meet with existing room. Allowance made for carpeting.
- Must be level.

Windows

- Andersen No. 3456 for south wall.
- Replace existing window with Andersen No. 2832 and fir in. (See Fig. 4-22. *Note:* Size difference of windows.)
- Windows to be vinyl-coated and screened as per existing.
- Shutters on south window wall to match existing.

Door

- Interior with trim as per existing, mounted and operating.
- Size, 32″ × 6′8″; solid, flush, walnut with brass lock and hardware.
- Dining room to be left in excellent condition.

Permits

- To be obtained by the contractor.

Walls

- Studs, 2 × 4 kiln-dried (construction or better), 16 inches O.C.

- Double or better headers at windows and door and double at corners where necessary.

Roof

- Composition shingles as per existing, black of same weight as existing.
- Felt paper under shingles and attached to ⅝-inch exterior plywood (A—D) sheathing.

Rafters

- Kiln-dried 2 × 4 truss type as per existing, 16 inches O.C.
- Roof type to be shown in attached drawings. Check with existing garage to determine type of construction if necessary.

To add any information you may have left out, you can make drawings illustrating what you need. Figure 4-23 shows a cornice detail that ensures there is no misunderstanding of what the overhang is. Other details are also present in this drawing. The scale of the cornice is 1½″ = 1′0″.

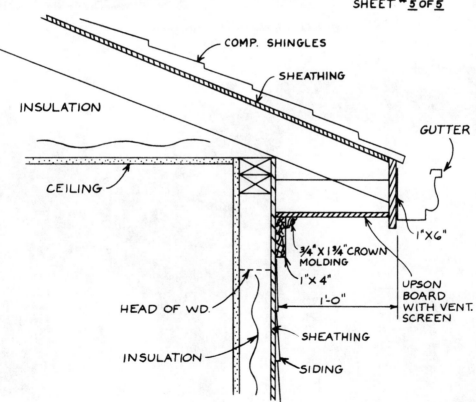

SHEET #5 OF 5

CORNICE DETAIL

Fig. 4-23 *Cornice detail.*

The contractor is protected when the details of work to be done are spelled out in this way. There is little or no room for argument if things are written out. A properly drawn contract between contractor and owner should also be executed to make sure both parties understand the financial arrangements. Figures 4-24 and 4-25 show the completed addition as it looks with landscaping added.

CREATING NEW STRUCTURES

A storage building can take the shape of any number of structures. In most instances you want to make it resemble some of the features of the home nearby.

Small storage sheds are available in precut packages from local lumber yards. Figure 4-26 shows three

Fig. 4-24 *Side elevation in finished form.*

Fig. 4-25 *Rear elevation in finished form.*

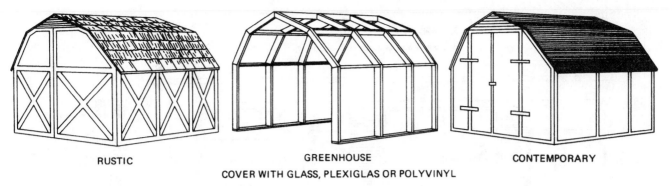

RUSTIC

GREENHOUSE
COVER WITH GLASS, PLEXIGLAS OR POLYVINYL

CONTEMPORARY

Fig. 4-26 *Various designs for storage facilities.* (TECO)

versions of the same package. It can be varied to meet the requirements of the buyer. Lawnmowers, bicycles, or almost anything can be stored in a structure of this type.

All you need is a slab to anchor the building permanently. In some instances you may not want to anchor it, so you just drive stakes in the ground and nail the soleplate to the stakes. Most of the features are simple to alter if you want to change the design.

Before you choose any storage facility, you should know just how you plan to use it. This will determine the type of structure. It will determine the size and in some cases the shape. The greenhouse in Fig. 4-26 is nothing more than the rustic or contemporary shed with the siding left off. The frame is covered with glass, Plexiglas, or polyvinyl according to your taste or pocketbook.

Custom-Built Storage Shed

In some cases it is desirable to design your own storage shed. The example used here, shown in Fig. 4-27, was designed to hold lawnmowers and yard equipment. It was designed for easy access to the inside. Take a look at the overhead garage door in the rear of the shed. The 36-inch door was used along with the window to make it more appealing to the eye. It also has a hip roof to match the house to which it is a companion. See Fig. 4-28.

The design is 10 feet × 15 feet long. See Fig. 4-29. That produces a 2-to-3 ratio which produces a pleasing appearance. A 3-to-4 ratio is also very common. The concrete slab was placed over a bed of crushed rocks and anchored by bolts embedded in the concrete slab. A 9-foot-wide and 7-foot-long slab is tapered down from the floor to the yard. This allows rider lawnmowers to be driven into the shed. The outside pad also serves as a service center.

Wires serving the structure are buried underground and brought up through a piece of conduit. They enter the building near the small door. There is only one window,

Fig. 4-27 *View of one end of a storage shed.*

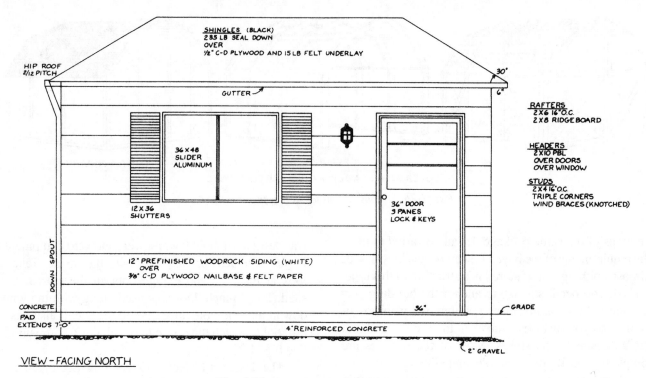

Fig. 4-28 *View of the finished storage shed.*

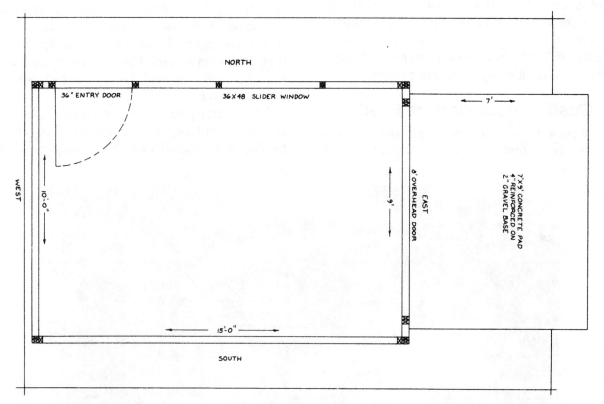

Fig. 4-29 *Outside dimension details.*

so the wall space can be used to hang yard tools. The downspouts empty into the beds surrounding the structure to water the evergreens. An automatic light switch turns on both lights at sundown and off again at dawn.

The overhead door faces the rear of the property. This produced some interesting comments from the concrete installers, who thought the garage had been turned around by mistake. It does resemble a one-car garage, but it is specifically designed for the storage of yard work equipment.

Don't forget to get a building permit. In some areas, even a tool shed requires a building permit. This doesn't necessarily mean it goes on the tax rolls, but does call for a number of inspections by the building inspector, which helps protect both builder and owner,

Buildings for storage take all shapes. They may be garages or barns. The design of a new structure should be carefully chosen to harmonize with the rest of the buildings on the property.

Building Codes

Building codes are laws written to make sure buildings are properly constructed. They are for the benefit of the buyer. They also benefit all the people in a community. If an expensive house is built next to a very inexpensive one it lowers the property value of the expensive house. Codes are rules which direct people who build homes. Codes say what can and cannot be done with a particular piece of land. Some land is hard to build on. It may have special surface problems. There may be mines underneath. There are all kinds of things which should be looked into before building.

If a single-family house is to be built, the building codes determine the location, materials, and type of construction. Codes are written for the protection of individuals and the community. In areas where there are no codes, there have been fires, collapsed roofs, and damage from storms.

Figure 4-30 shows inspections needed for a building permit. This is one way the community has to check on building. You have to obtain a permit to add on to a house. It is necessary to get a permit if you build a new house. Figure 4-31 shows the permit for a tool shed. One reasons for requiring a building permit is to let the tax assessor know so that the property value can be changed.

Note in Fig. 4-31 how the application for a building permit is filled out. Note also the possibilities to be checked off. The town board is required to sign, since it is responsible to the community for what is built and where.

In Fig. 4-30, look at the number of inspections required before the building is approved. Each inspection is made by the proper inspector. This way the person who buys the house or property is protected.

The certificate of occupancy is shown in Fig. 4-32. This is required before a person can move into the house. The certificate is given with the owner in mind. It means the house has been checked for safety hazards before anyone is allowed to live in it.

NOTICE
INSTRUCTIONS AND REQUIRED INSPECTIONS

A. Builder's name, phone number, property house number, and building permit number must be displayed on all construction or building projects.

B. A reasonable means of ingress must be provided to each structure and each floor. (Planks or other; and ladders or stairs from floor-to-floor.)

C. The following inspections are mandatory on all construction within the Town of Amherst:
   1. FIRST WALL: When footing is ready to be poured, including trench pour.
   2. PRE-LATH: Before insulating and after ALL electrical, telephone, plumbing, and heating rough work is complete, including metal gas vent and range-hood exhaust duct.
   3. DRAIN TILE: Subfloor and/or footing drain tile, before pouring of concrete floor.
   4. FINAL: Make application for Certificate of Occupancy, file copy of survey with application. Everything must be completed. All work must be finished, including sidewalks and drive aprons.) Each structure must have a house number displayed and visible from the street.

NOTE: An approved set of building plans and plot plans shall be made available to the Building Inspector at the above four inspections.

Call the Building Department 24 hours before you are ready for inspection.
Phone 555-6200, Ext. 42, 43, or 49

Fig. 4-30 Notice of the inspections needed when a building permit is required for a house.

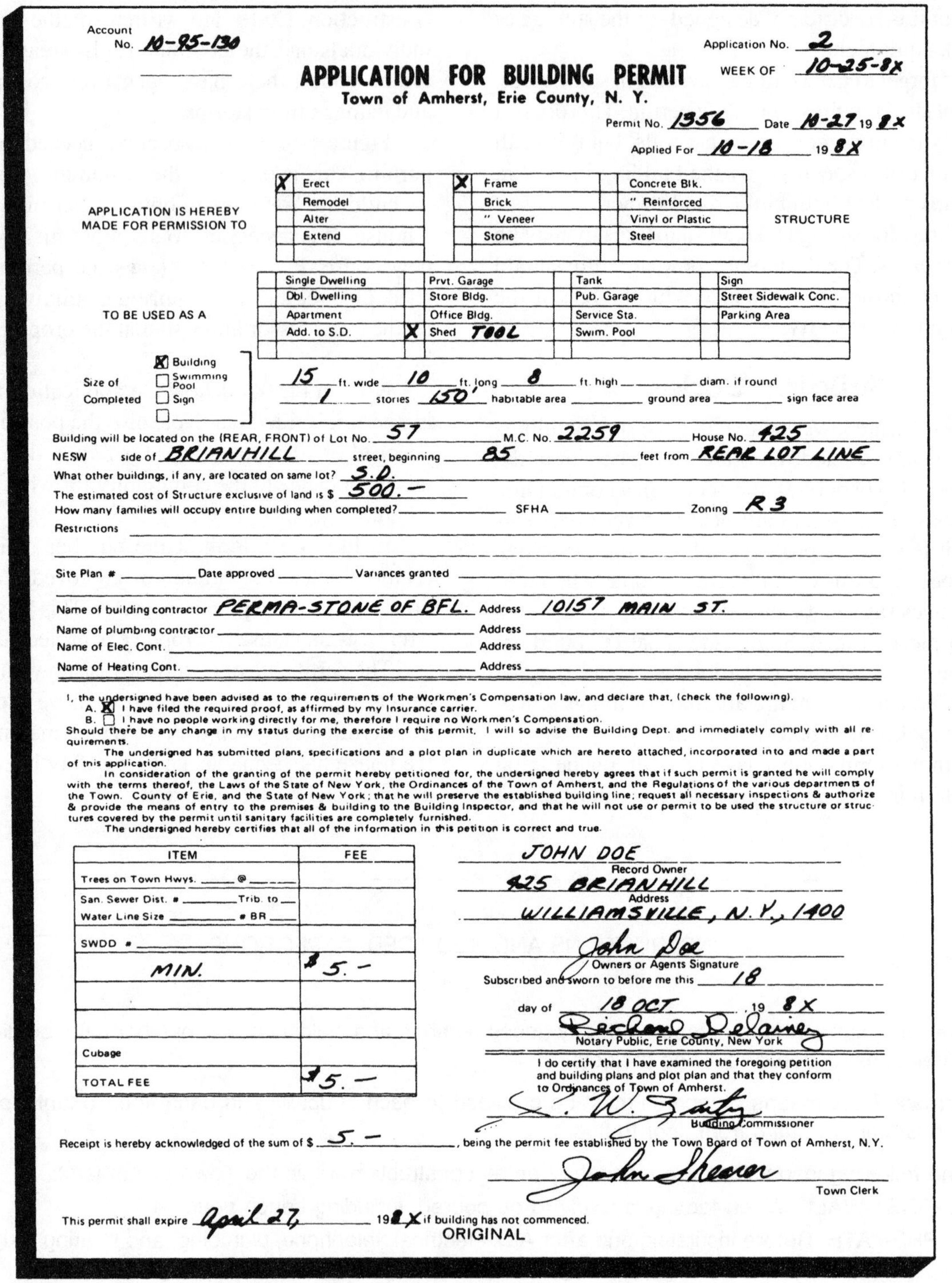

Fig. 4-31 *Application for building permit.*

Community Planning and Zoning

Zoning laws or codes are designed to regulate areas to be used for building. Different types of buildings can be placed in different areas for the benefit of everyone. Shopping areas are needed near where people live. Working areas or zones are needed to supply work for people. In most cases, we like to keep living and working areas separated. This is so because of the nature of the two types of buildings. Some people don't like to live near a factory with its smoke and noise.

Living near a sewage treatment plant can also be very unpleasant. Certain types of buildings should be near one another and away from living zones.

Some areas are designated as industrial. Others are designated as commercial. Still others are marked for use as residential area. Residential means homes. Homes may be single houses or apartments. An industrial building cannot be built in a residential area. All over the United States there are regional planning boards. They decide which areas can be used for what. Master plans are made for communities. Master plans designate where various types of buildings are located.

Community plans also include maps and areas outlined as to types of buildings. Parks are also designated in a community plan. Streets are given names, and maps are drawn for developments. A development means land to be developed for housing or other use. Figure 4-33 shows a typical plan for a residential development. Note

TOWN OF AMHERST

CERTIFICATE OF OCCUPANCY

Date June 25, 19XX No. 1800

This Certifies that the building located at and known as

125 Briarhill

(NO.) (STREET)

Sub Lot No. 57 Map Cover No. 2259

and used as a single-family dwelling with private garage

(KIND OF OCCUPANCY)

has been inspected and the use thereof conforms to the Amherst Zoning Ordinance, and other controlling laws and is hereby:

( ) Approved

(x) Approved subject to the planting of the required number of trees not later than Oct. 15, 19XX.

Any alteration of the property or change in the use voids this certificate and a new certificate will be required.

COMMISSIONER OF BUILDING

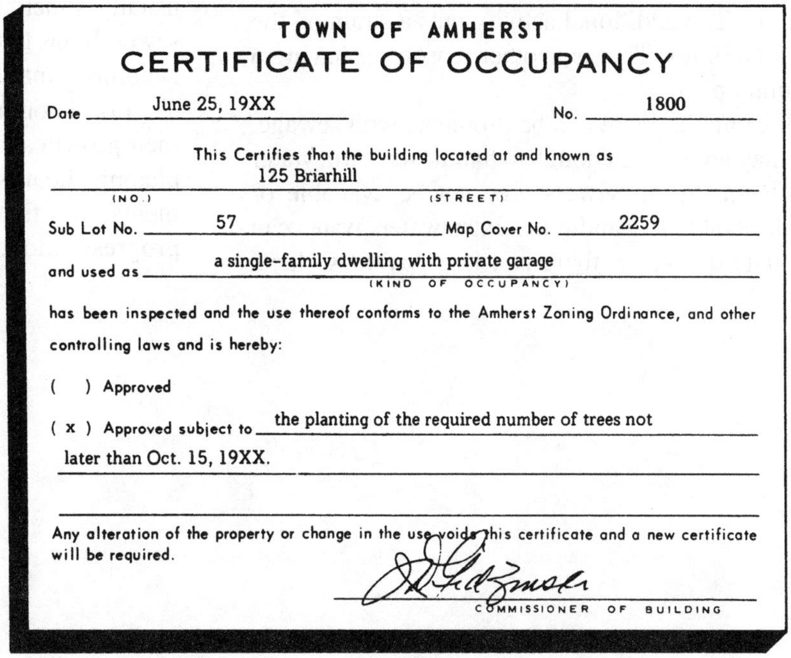

Fig. 4-32 *Certificate of occupancy.* *(Courtesy of Town of Amherst, NY)*

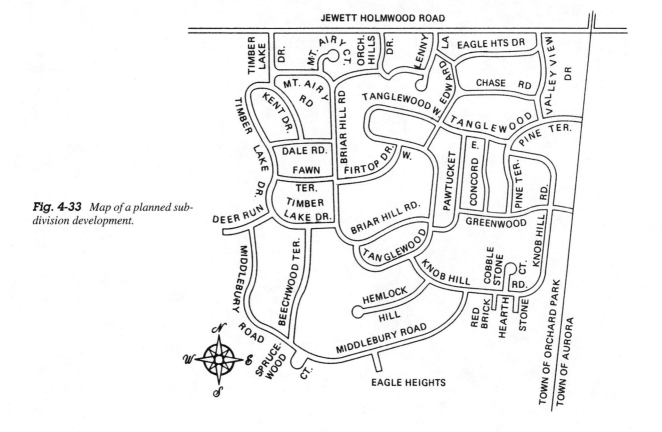

Fig. 4-33 *Map of a planned subdivision development.*

how the streets are laid out. They are not straight rows. This has a tendency to slow down traffic. The safety of children is important in a residential area. It is best to have residential areas off main traveled roads or streets.

Overbuilding

Where there has been residential, commercial, and industrial overbuilding, the utilities have not been able to keep up. The additional activity puts a strain on the existing facilities. That is another reason for having a community plan.

For example, there can be problems with sewage. Plants may not be big enough to handle the extra water and effluent. Storm sewers may not be available or may not be able to handle the extra water. Water can accumulate quickly if there is a large paved surface.

This water has to be drained fast during a rainstorm. If it is not, flooding of streets and houses causes damage. Some areas are just too flat to drain properly. It takes a lot of money to build sewers.

Sewers are of two types. The sanitary sewer takes the fluid and solids from toilets and garbage disposals. This is processed through a plant before the water is returned to a nearby river or creek. The storm sewer is usually much larger in diameter than the sanitary sewer. It has to take large volumes of water. The water is dumped into a river or creek without processing.

Local communities can have a hand in controlling their growth and problems. They can form community planning boards. These boards enforce zoning requirements, and the development of the community can progress smoothly.

5
CHAPTER

Painting and Wallpapering

PAINT SELECTION

Paints come in a wide variety and vary considerably in cost. In most cases, the more expensive paints are more costly because they have more pigments, additives, and solids. The pigments provide the color, so if there are more pigments in the paint, then the paint will have better coverage. Therefore, it should take fewer coats of paint to cover a surface when you use a more expensive paint. Additives in paint improve the performance of paint in many ways. Some of these benefits include better color retention and durability, increased mildew resistance, improved adhesion qualities, ease of application, and better resistance to dirt and acid rain. Typically, expensive paints contain 40% or more solids while less expensive paints contain 30% or less. The solids are the actual paint while the remainder of the paint is a liquid. The liquid is usually water for latex paints and some form of oil for oil-based enamel paints.

Although higher-quality paints may be more expensive, their costs are justified because the paint covers better and you will end up purchasing less paint. In addition, the paint should last longer, so you will not have to repaint as often. Labor costs associated with painting are much more expensive than the paint itself. Therefore, the less often you have to paint, the more money and time you will save in the future.

Water-Based Paints versus Oil-Based Paints

Water-based paints are the latex paints. There are so many variations of paint nowadays that is often difficult to tell a water-based paint from an oil-based paint when you are reading the label. The easiest way to tell the difference is to read the directions and look for what substance is used to thin the paint or clean the brushes. If mineral spirits or paint thinner is listed, then the paint is oil-based. Water-based paints will simply require warm water and soap to clean the brushes.

Because of environmental concerns, it would be best if all paints could be water-based. Oil-based paints require the use of solvents such as mineral spirits and paint thinner that are washed down the drain during the cleaning of brushes and other tools. These chemicals then have to be cleaned out of the tap water. Because of this, it is always best to use a water-based or latex paint whenever possible.

In most cases, a latex paint can be used for almost every situation; however, there are a few instances when oil-based paints clearly outperform water-based paints. These instances include using an oil-based primer or paint to cover metal surfaces. In most cases, when a water-based paint is applied over a nail or other metal surface, rust will bleed through within a few hours. In some instances, it may take a few months for the brownish-red spot of rust to bleed through your paint, which becomes more of a problem when all your furniture has been put back in its proper place.

Oil-based paints are preferred by most professional painters for painting trim, doors, cabinets, and other surfaces that are subjected to a lot of use. These paints are preferred because oil-based paints are usually glossier than latex gloss paints. In addition, most oil-based paints do not tend to harden as quickly underneath the surface, so they are not as brittle and do not chip as easily.

SURFACE PREPARATION

One of the most important parts of a successful paint job is surface preparation. If the surface is not properly treated before painting, then the paint will not adhere and the job will have to be redone. Most people do not have time to paint, let alone work on a paint job a second time. The first thing to remember is to read the directions on the can of paint. In most cases, the paint manufacturer will list in the directions the preferred method to prepare the work surface before applying the paint. If these steps are not listed on the can of paint, then usually there are material data sheets or application instruction sheets available close to the display of paint.

Typically, the following rules will apply to most paints available to the consumer:

1. Paint will not stick to a shiny (glossy) surface, so the surface must be sanded, preferably with a 220-grit sandpaper. In hard-to-reach places, a medium-grit foam sanding pad or scuffing pad will scratch the surface so that paint will adhere (Fig. 5-1). Remember, the entire surface should look dull, or else the paint will not stick or will peel off after a few weeks and/or months.

2. Rust must be thoroughly treated before the application of paint. Covering rust with paint will trap moisture and cause it to bubble up and spread faster than if it were not painted. The best method for removing rust is sandblasting (Fig. 5-2). In this method, sand is forced through the orifice of a gun at 80 pounds per square inch (psi) or higher. The air pressure forces the sand to knock off the rust. Additionally, the air pressure blows all the rust off the surface.

   Follow all the safety rules for sandblasting equipment because the sand is traveling at a high velocity. Therefore, it can remove your skin just as easily as it can remove rust and paint. Make sure all of your body is covered with a thick material, such

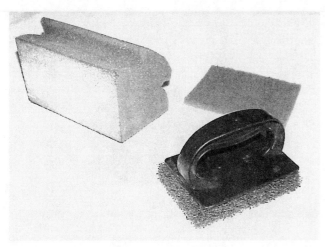

Fig. 5-1 *Plastic sanding pads.*

Fig. 5-3 *Paint spray respirator.*

Fig. 5-2 *Sandblasting equipment and clothing.*

as gloves for your hands and a jacket for your body. A clear face shield should also be worn to deflect the sand particles from your face. Safety goggles or glasses should be worn under the safety shield because the sand has a tendency to bounce all over the place. Your head should also be covered with some type of hat. A respirator (Fig. 5-3) should be worn because the very fine dust and sand particles are not good for your lungs. Most equipment rental stores have a sandblasting suit available with the sandblaster. You may have to purchase an additional face shield for the suit because the shields get scratched from the sand when sandblasting.

Grinding or sanding rust, on the other hand, forces rust back into the surface. However, since sandblasting is very dusty and makes quite a mess, sanding or grinding rust may be the only alternative. In this case, make sure the metal surface is treated with a metal conditioner or some type of phosphoric acid, which will neutralize the rust after grinding. Follow the directions on the metal conditioner, and make sure the metal surface is completely dry before you apply paint. Trapping in moisture will only cause the surface to rust in a few months. Use a blow dryer to dry the metal if you are in a hurry. The air pressure from the blow dryer will also force any moisture out of any cracks or surface imperfections in the metal.

3. Chips in the paint of trim and other shiny surfaces must be filled or feathered in by sanding so they do not stand out. High-gloss paint does not fill defects; on the contrary, all imperfections become more visible to the naked eye.

4. Flat or matte surfaces do not have to be sanded before painting. This is why most walls, ceiling, and other large surfaces are usually painted with flat paints. Some people prefer to use a semigloss paint in kitchens and bathrooms; however, the walls must be sanded or scuffed up before you repaint them. If the walls are not sanded, the paint will easily come off with use. Semigloss paints are used so spills or splatter marks on the walls can easily be cleaned with a wet rag. However, walls can just as easily be cleaned if expensive flat or satin wall paint is used. A shiny wall will also bring out the wall surface imperfections. Therefore, greater surface preparation will be needed before painting.

5. Once a wall is sanded or scuffed up, all the dust must be washed and wiped off. Before painting, all surfaces must be thoroughly cleaned of dirt, oil, grease, wax, or any other substance that paint will not adhere to. Metal or wooden surfaces that will be stained, primed, or clear-coated with a finish should be wiped off with a tack rag. A tack rag is a sticky, woven cloth that comes in a sealed wrapper. Tack rags pick up very fine dust particles that would be seen in a high-gloss finish.

PAINTING

In most cases, the high-gloss trim areas are painted first. In this way, masking of walls will not be needed because the wall paint will be applied over it later. The walls are painted after the trim because excess paint can easily be wiped off with a damp rag from a shiny surface. Masking tape will also adhere better to a smooth, shiny surface than dry wall that may have some type of rough decorative finish.

When purchasing paint by the gallon, you may want to consider paint that is sold in a plastic container (Fig. 5-4). Most metal paint cans rust around the rim after a few years, so if you want to reopen the can to touch up a paint spot, you will end up breaking the can or getting rust particles in your paint. An alternative to this is to purchase custom metal lids that have a plastic spout built into them (Fig. 5-5). These lids cost less than $2, which is cheaper then buying another gallon of paint. In addition, paint manufacturers change their colors quite often, so it is very difficult to find the same color later.

Fig. 5-4 *One-gallon house paint containers.* (Courtesy Dutch Boy Paints.)

Brush

Trim is usually painted with a high-quality brush that has a fine bristle pattern. In most cases, the bristles are cut so that they form a tapered straightedge or an angled edge (Fig. 5-6). The angled edge has the advantage of

Fig. 5-5 *Custom metal lid with built-in plastic spout.*

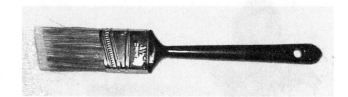
Fig. 5-6 *Paint brush with tapered bristle pattern.*

allowing you to apply paint in corners and other hard-to-reach places (Fig. 5-7). Sometimes inexpensive chip brushes make excellent paint brushes because of their fine bristles (Fig. 5-8). Be prepared to pick out some of the bristles when you first start painting. Do *not* use foam brushes when you apply oil-based paint because the plastic used to make most of these brushes dissolves in lacquer thinner and/or paint thinner.

Fig. 5-7 *Tapered brush used to paint a corner.*

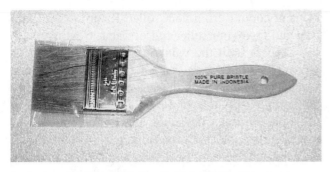

Fig. 5-8 *Inexpensive chip brush.*

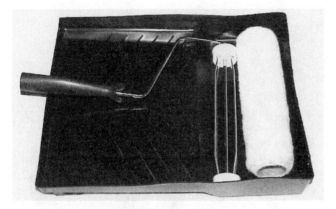

Fig. 5-9 *Paint tray with roller.*

When you paint with a brush, dip the bristles only one-third of the way into the paint, and brush the excess onto the side of the can or tray. Too much paint will drip and run down your painting surface. Make long strokes the entire length of the surface being painted. In addition, try to place as much paint on the surface as possible without its running. A thick coat of paint will level out and hide the brush marks. Always look back on what you previously painted while you are painting, because sometimes it takes a few minutes for runs and sags to start. It is better to touch up a run while the paint is still wet than to have to sand it and repaint it later.

Thoroughly clean a brush in the correct solvent several times, and then wash it out with soap and warm water. If even a thin coat of paint is allowed to dry on the bristles of the brush, it will become hard and leave unsightly brush marks in your work.

Roller versus Airless Sprayer

The walls and other large surfaces are painted using a roller and a paint tray (Fig. 5-9). A power airless sprayer (Fig. 5-10) can be used; however, the overspray will splatter with paint items that are not masked off. A roller is probably the best tool for repainting one

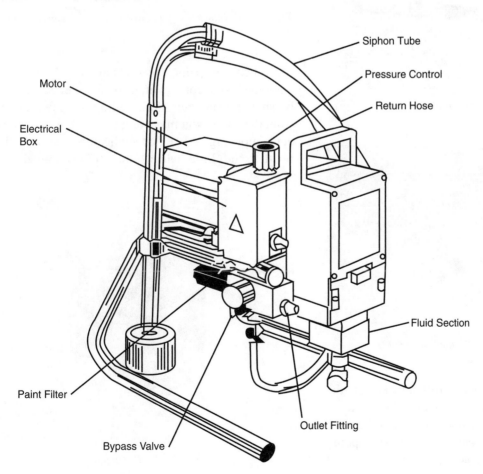

Motor

Electrical Box

Siphon Tube

Pressure Control

Return Hose

Fig. 5-10 *Airless sprayer.*

Fluid Section

Paint Filter

Outlet Fitting

Bypass Valve

or two rooms. There are a wide variety of rollers; however, if you are just painting a smooth, flat wall with a water-based latex paint, then usually the thin inexpensive rollers will suffice. If the surface to be painted is coarser, then a roller with a thicker surface covering should be used. Do not use the least expensive rollers with oil-based paints because the woven material will come off in clumps and ruin your painting surface. Read the labels carefully before you purchase a roller for your particular job.

When you paint a room, it is best to paint the ceiling first, so if you drip paint, it will splatter only on the walls which will be painted later. Most paint roller handles are threaded so a broomstick handle can be screwed into them (Fig. 5-11). With the increased length of the handle, a roller can easily reach the ceiling. Before you paint, make sure the paint tray has a disposable liner (Fig. 5-12) in it or is lined with aluminum foil. This makes cleanup after the job much easier, and the remaining paint is not washed down the drain to pollute our tap water.

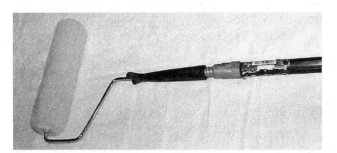

Fig. 5-11 *Roller with broomstick extension.*

Fig. 5-12 *Disposable paint tray liner.*

To evenly apply paint with a roller, make sure the entire surface of the roller is covered by paint by rolling it up and down the inclined surface of the paint tray. Start on one side of a room and keep overlapping your paint strokes until you no longer see a solid coat

of paint coming from your roller. Return to the paint tray and load your roller full of paint. Continue with this process until the painting surface is finished. If you are repainting a wall with the same color paint and you rolled the paint on thick, then you will probably not have to apply a second coat of paint. Remember to use a primer sealer if you are applying a lighter color over a darker color or if the painting surface has never been painted before. Unsealed brick, paneling, wood, or other coarse materials will never seal with paint and will fade or bleed (Fig. 5-13) within a few weeks or months.

Fig. 5-13 *Unprimed paint with wood sap bleeding through paint.*

When you paint walls, always take a brush and paint the inside corners first, so the roller can level the brush marks out later. If the ceiling is a different color from the walls, you may want to consider using a pad applicator (Fig. 5-14). Pad applicators have very short bristles and little wheels that follow an edge, so

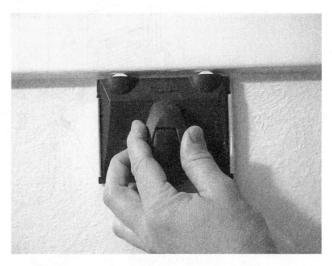

Fig. 5-14 *Pad applicator.*

you get a nice straight line of paint. Make sure you keep the edge and wheels clean, or you will get paint on the surface that you do not want to paint.

An airless sprayer should be considered when a large number of rooms must be painted. Airless sprayers typically have a spray pattern (fan) of 18 to 24 inches and can easily spray thick latex paint without it being thinned. Note that only sprayers with ⅜ horsepower or greater can spray large volumes of paint for a long time without any downtime.

A person with an airless sprayer can paint an entire house of approximately 2000 square feet in 8 hours or less. If the sprayer is used to apply paint to the exterior of the house on a hot day, the spray nozzle will tend to clog when not in use. To prevent this, keep the spray gun submerged in a pail of water until you start spraying again. The paint inside the gun will not have a chance to dry and clog the spray gun. This technique is only suited for airless sprayers in which the spray gun is separate from the electric or gasoline motor. An airless sprayer must be thoroughly cleaned by running water through the gun instead of paint for approximately 5 to 10 minutes immediately after use. A rust inhibitor and lubricant should be added to the water during the last minute of cleaning. This product is available wherever airless sprayers are purchased or rented.

Remember to wear safety glasses, disposable plastic gloves (Fig. 5-15), a hat, and clothing to cover your body because spray guns produce overspray that will splatter on your body. Airless sprayers force paint through the nozzle at high pressure so never point the nozzle of the gun at your face and spray it while cleaning.

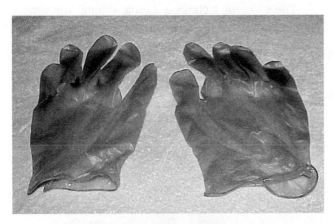

Fig. 5-15 *Disposable plastic gloves.*

Siphon Feed Sprayer

Conventional siphon feed spray guns (Fig. 5-16) that require the use of an air compressor are used for spraying thinned materials such as polyurethane enamel for

Fig. 5-16 *Conventional siphon feed spray gun.*

cabinets. The spray pattern for these guns is usually 8 inches or less. These spray guns break the paint into finer droplets, which creates a smoother finish with less of an "orange peel" appearance. Speed is traded for appearance. Typically, surfaces with a high-gloss finish are painted with a siphon feed sprayer.

A respirator must be worn whenever you use a spray gun. In addition, disposable plastic gloves will save you a lot of time cleaning your hands. Many oil-based products are toxic so you should always try to avoid contact with your skin. Thinners will dry out your hands, so make sure you wear gloves at all times.

Siphon feed spray guns must be cleaned immediately after use with the proper thinner. The spray cup or container for the gun must also be cleaned. Once the cup is completely clean, fill it again half full with the proper thinner and spray the gun until only the thinner is spraying out of the gun. A spray gun that is not thoroughly cleaned will never spray evenly and smoothly and will then have to be rebuilt.

Spraying with a Spray Gun

Before you spray, make sure the paint is thoroughly mixed with a wooden stirring stick. Read the directions on the paint can because some polyurethane paints should not be shook or stirred because that creates bubbles in the paint. Always use a strainer when you pour the paint into the cup or container of the spray gun. Any contamination can clog the spray gun and/or ruin the appearance of the painted surface.

Spraying with a spray gun is more difficult than you would think. The reason is that you have to keep an equal distance from your painting surface as you make a complete pass with the gun. It is human nature to bend your arm and not your wrist, so you end up making a curved pass at your painting surface (Fig. 5-17). If this happens, wherever you came closest to the painting surface (which would be in the middle of your pass) the paint is put on thicker and will therefore run or sag. To prevent this, bend your wrist as you spray from left to right, always maintaining an equal distance of 6 to 12 inches depending upon the paint and how much it has been thinned (Fig. 5-18). The trick is to practice your paint stroke on a sample board made out of the same material that you are going to paint. It is much better to make a mistake on a sample board than to have to sand the runs and sags out of your painting surface later.

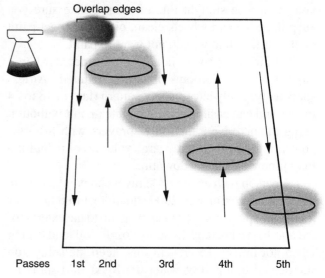

Overlap edges

Passes   1st   2nd   3rd   4th   5th

Fig. 5-19   Spray stroke overlap.

of paint to spray a surface evenly and smoothly. The second coat of paint should also be sprayed vertically instead of horizontally, therefore covering all parts of the painting surface.

PAINTING EFFECTS

Wallpaper allows you the ability to apply any pattern to your wall; however, it is often very expensive and difficult to apply. In addition, if it gets torn or damaged in any way, it is almost impossible to repair. Over the years, wallpaper seams and border treatments can also start lifting, and the paper finally starts to fade. An alternative to wallpaper is to use the following painting effects that, in many instances, emulate wallpaper.

The *sponging* technique is quite simple. You simply paint a wall; once it dries, you go back over it with a sponge dipped in another color of paint (Fig. 5-20). The trick is to keep an evenly spaced pattern while sponging so that some marks do not seem closer than others.

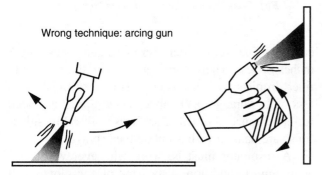

Wrong technique: arcing gun

Fig. 5-17   Incorrect curved pass spray stroke.

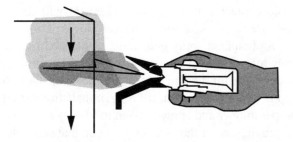

Fig. 5-18   Equal-distance spray stroke.

When painting a surface, you should always overlap your paint strokes approximately 50 percent (Fig. 5-19). High-gloss paint is harder to apply because if you apply it too thickly, then where it appears shiny it often runs. However, if you do not apply it too thickly, then the surface looks dull and rough. In most cases, it is best to apply a thin, dull coat; let it dry; and then apply a thicker, shiny coat of paint. It always takes two coats

Fig. 5-20   Sponge technique.

It is best to overlap the sponge marks so the pattern stays consistent. Another trick to sponging is to find similar shades of color that are next to each other on a color sample sheet that you can get at any paint display. An uneven sponge pattern will not stand out as much if the paint colors are similar. First the darker color should be applied, and then the lighter color should be sponged on top of it. In this manner, uneven sponge patterns will not be as noticeable. Store-bought sponges with various shapes can be purchased for a more dramatic effect (Fig. 5-21).

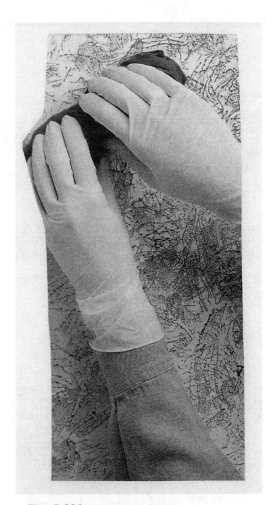

Fig. 5-22A *Ragging technique.* *(Courtesy Lowe's)*

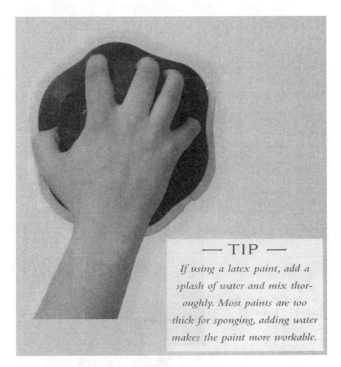

— TIP —

If using a latex paint, add a splash of water and mix thoroughly. Most paints are too thick for sponging, adding water makes the paint more workable.

Fig. 5-21 *Sponge with precut design.* *(Courtesy Lowe's)*

Ragging is a technique that is very similar to sponging. Instead of applying the second coat of paint with a sponge, you use an old, rolled up piece of cotton underwear or similar material. You take the rag, submerge it in paint, and then roll it over the wall at different angles. Make sure you are wearing disposable plastic gloves, or you will be washing your hands for a few hours. This technique is quite slow, so you may want to rag only smaller rooms such as bathrooms and foyers. An example of this technique is shown in Fig. 5-22A and B.

In the *dragging* technique, the wall is painted with a base coat, and then once it has dried overnight, a glaze (different color) is rolled on in 4- to 6-feet sections. Immediately after applying the glaze, you hold a clean, wide paintbrush or an even wider wallpaper smoothing brush parallel to the wall and then drag it evenly down the entire wall (Fig. 5-23A). The streaking effect is illustrated in Fig. 5-23B.

Fig. 5-22B *Ragging technique.*

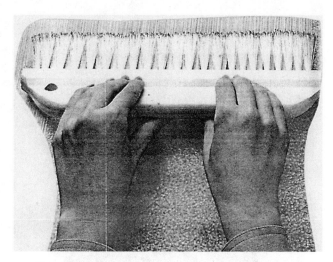

Fig. 5-23A *Dragging technique with wallpaper brush.*

Fig. 5-23B *Dragging technique with paintbrush.*

Fig. 5-24 *Stippling technique.*

Fig. 5-25 *Roller and edge stamp for rolling technique.* (Courtesy Pattern Magic)

The *stippling* technique is done in the same manner as the dragging technique except that you would use a stippling brush to make a textured pattern, as seen in Fig 5-24. Special stippling brushes can be purchased where paint glazes are sold.

The *rolling* technique is done by purchasing special rollers that have a wide variety of patterns on them. Since a round roller cannot fit into an interior corner, little square-edge stamping tools are also included in the kits (Fig. 5- 25). Using a laser level or popping a chalk line would be useful to keep the pattern straight. In most cases, the rollers follow the top edge of a wall or wainscot and are used as a border.

More painting effects seem to appear each day, so visit your local paint supply stores before you decide upon an effect.

MILDEW

Harmful mildew growth (black spots) is a major concern because if it spreads throughout your home, it can be very harmful to human beings and pets (refer to Fig. 5-26).

Fig. 5-26 *Mildew on ceiling in bathroom.*

Fig. 5-27 *Tree shading a house.*

Removal of mildew is similar to asbestos removal and is very costly. In fact, most insurance companies will not cover mildew repairs unless you pay a higher insurance premium. Mildew growth can only occur in damp places, so the removal of the mildew and moisture must be done before repainting.

Mildew usually occurs in bathrooms and kitchens or wherever there is some sort of water supply. Typically, bathrooms with showers have the most problems with mildew. The steam from showers will absorb into the wallpaper or sheetrock, and mildew will grow beneath the surface. Adequate ventilation will easily prevent the spread of mildew. In most cases when mildew is present, bathrooms do not have ceiling ventilation fans or the fans are too small for the job. In addition, shower stalls should be wiped down with a dry towel after every use. The bathroom door should also be left open during the day so the room can dry out.

The best way to treat mildew that is already present is to mix 1 part of bleach with 3 parts water and spray the mixture on the surface. Make sure you wear disposable plastic gloves while cleaning because mildew and bleach are harmful substances. Once the mildew has been removed, spray the surface for at least another 30 minutes with bleach to kill any remaining mildew particles that are not noticeable to the eye. Thoroughly air-dry the surface for a day, and then use a blow dryer to make sure the surface is dry and clean. The surface is now ready to be primed and painted.

If you are having problems with mildew growth on an exterior side of your house, it is probably due to the fact that a tree is shading it and no sunlight is getting through to dry the surface. In this case, you should trim your tree (Fig. 5- 27) and then clean the house with bleach as previously mentioned. In addition, there are additives available that you can mix with your paint to prevent the growth of mildew. These additives are available at most stores that sell house paint.

It is important to remove any signs of mildew, especially if you want to sell your home in the near future. In many states, a realtor cannot legally sell a home that shows any sign of mildew growth. Moreover, if you wait too long to treat mildew, it can spread into the interior studs of your home and your whole house may have to be torn apart or, worse yet, condemned. Remember, if you do not have insurance coverage for mildew, then you would be responsible for the entire cost of your home.

LEAD-BASED PAINTS

Many homes built before 1978 have lead-based paint. The federal government banned lead-based paint in 1978, so homes built after that date should not contain any lead in the paint. Several states banned lead-based paint several years before 1978, so you may want to check with your local building inspector. Lead-based paint is usually not a hazard if it is in good condition and is not in an area of high use, such as a window or door. Lead-based paint becomes a problem when it flakes off in chips or when it is sanded and turns into dust. These chips or dust can be ingested into the human body and cause the following:

1. Damage to the brain and nervous system
2. Behavior and learning problems (such as hyperactivity)
3. Slowed growth
4. Hearing problems
5. Headaches
6. Difficulties during pregnancy

7. Reproductive problems in men and women
8. High blood pressure
9. Digestive problems
10. Nerve disorders
11. Memory and concentration problems
12. Muscle and joint pain

Lead-based paint can be identified by a trained, certified professional who will use a range of reliable methods to check your home, such as a portable X-ray fluorescence machine and laboratory samples of paint, dust, and soil. You can call 1-800-424-LEAD for a list of certified contacts in your area. There are home test kits available for determining lead content in paint; however, they are not always accurate. For more information concerning lead disclosure and reducing your exposure to lead, please refer to Figs. 5-28 and 5-29.

Are You Planning To Buy, Rent, or Renovate a Home Built Before 1978?

Many houses and apartments built before 1978 have paint that contains high levels of lead (called lead-based paint). Lead from paint, chips, and dust can pose serious health hazards if not taken care of properly.

Federal law requires that individuals receive certain information before renting, buying, or renovating pre-1978 housing:

LANDLORDS have to disclose known information on lead-based paint and lead-based paint hazards before leases take effect. Leases must include a disclosure form about lead-based paint.

SELLERS have to disclose known information on lead-based paint and lead-based paint hazards before selling a house. Sales contracts must include a disclosure form about lead-based paint. Buyers have up to 10 days to check for lead.

RENOVATORS have to give you this pamphlet before starting work.

IF YOU WANT MORE INFORMATION on these requirements, call the National Lead Information Center at **1-800-424-LEAD (424-5323).**

This document is in the public domain. It may be reproduced by an individual or organization without permission. Information provided in this booklet is based upon current scientific and technical understanding of the issues presented and is reflective of the jurisdictional boundaries established by the statutes governing the co-authoring agencies. Following the advice given will not necessarily provide complete protection in all situations or against all health hazards that can be caused by lead exposure.

Fig. 5-28 *Federal law regarding lead-based paint.*

What You Can Do Now To Protect Your Family

If you suspect that your house has lead hazards, you can take some immediate steps to reduce your family's risk:

◆ **If you rent, notify your landlord of peeling or chipping paint.**

◆ **Clean up paint chips immediately.**

◆ **Clean floors, window frames, window sills, and other surfaces weekly.** Use a mop or sponge with warm water and a general all-purpose cleaner or a cleaner made specifically for lead. REMEMBER: NEVER MIX AMMONIA AND BLEACH PRODUCTS TOGETHER SINCE THEY CAN FORM A DANGEROUS GAS.

◆ **Thoroughly rinse sponges and mop heads after cleaning dirty or dusty areas.**

◆ **Wash children's hands often, especially before they eat and before nap time and bed time.**

◆ **Keep play areas clean.** Wash bottles, pacifiers, toys, and stuffed animals regularly.

◆ **Keep children from chewing window sills or other painted surfaces.**

◆ **Clean or remove shoes before entering your home to avoid tracking in lead from soil.**

◆ **Make sure children eat nutritious, low-fat meals high in iron and calcium,** such as spinach and dairy products. Children with good diets absorb less lead.

Fig. 5-29 *How to protect against lead-based paints.*

PUTTING WALLPAPER ON WALLS

Wallpaper is widely used in buildings today. However, the "paper" may not be paper. It may be various types of vinyl plastic films. Also, a mixture of plastic and paper is common. In either case, the wall surface should be prepared. It should be smooth and free of holes or dents. As a rule, a single roll of wallpaper will cover about 30 square feet of wall area. A special wall sealer coat called *size* or *sizing* is used. This may be purchased premixed and ready to use. Powder types may be purchased and mixed by the worker. In either case, the sizing is painted on the walls and allowed to dry. As it dries, it seals the pores in the wall surface. This lets the paste or glue adhere properly to the paper.

A corner is chosen for a starting point. It should be close to a window or door. The width of the wallpaper is measured out from the corner. One inch is taken from the width of the wallpaper. Make a small mark at this distance. The mark should be made near the top of the wall. For a 27-inch roll, 26 inches is measured from the corner. A nail is driven near the ceiling for a chalk line. The chalk line is tied to this nail at the mark. The end of the chalk line is weighted near the floor. The line is allowed to hang free until it comes to a stop. When still, the line is held against the wall. The line is snapped against the wall. This will leave a vertical mark on the wall. See Fig. 5-30A.

Next, several strips of paper are cut for use. The distance from the floor to the ceiling is measured. The wallpaper is unrolled on the floor or a table. The pattern side is left showing. The distance from the floor to the ceiling is laid off, and 4 inches is added. This strip is cut, and several more are cut.

The first precut strip is laid on a flat surface. The pattern should be face up. The strip is checked for appearance and cuts or damage. Next, the strip is turned over so that the pattern is face down. The paste is applied with the brush. See Fig. 5-30B. The entire surface of the paper is covered. Next, the strip is folded

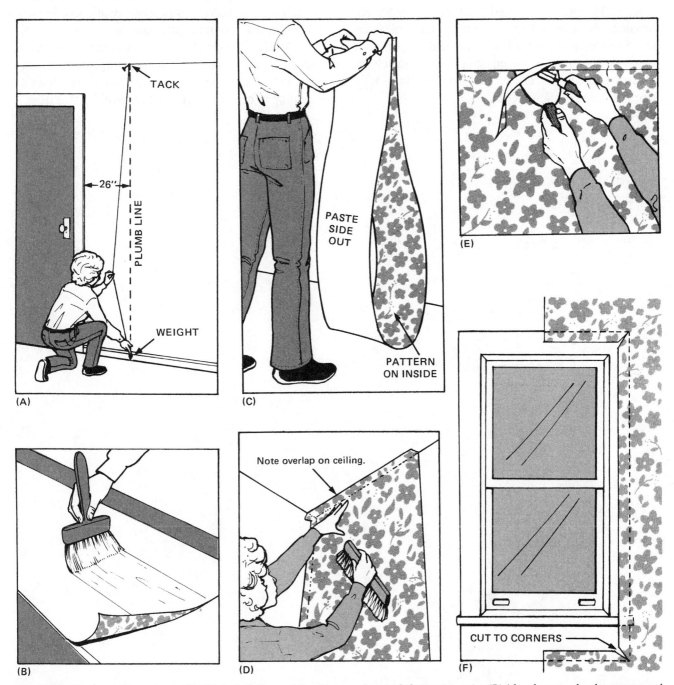

Fig. 5-30 *Putting wallpaper on walls. (A) A chalk line and plumb are used to mark the starting point. (B) After the paper has been measured and cut, paste is brushed on the back. (C) The bottom and top edges are used to carry the pasted wall covering. (D) Strips are lapped at the ceiling and brushed down. (E) Use a putty knife or straightedge as a guide to trim overlaps. (F) Cut corners diagonally at windows and doors.*

in the middle. The pattern surface is on the inside of the fold. This allows the worker to hold both the top and the bottom edges. See Fig. 5-30C.

The plumb line is used as a starting point. The first strip is applied at the ceiling. About 1 inch overlaps the ceiling. This will be cut off later. A stiff bristle brush is moved down the strip. See Fig. 5-30D. Take care that the edges are aligned with the chalk line. About 1 inch will extend around the corner. This will be trimmed away later. This is an allowance for an uneven corner.

The entire strip is then brushed to remove the air bubbles. The brush is moved from the center toward the edge. The second strip is prepared in the same manner. However, care is taken to be sure that the pattern is matched. The paper is moved up or down to match the pattern. The edge of the second strip should exactly touch the edge of the first piece.

Next, the second piece is folded down. The edge is exactly matched with the edge of the first piece. The edge should exactly touch the edge of the previous piece. Then, any bubbles are smoothed out. The process is repeated for each strip until the wall is finished. Then the wall is trimmed.

The extra overlap at the ceiling and at the base is cut. A razor blade or knife is used with a straightedge as a guide. See Fig. 5-30E.

A new line is plumbed at each new corner. The first strip on each corner is started even with the plumb line. This way each strip is properly aligned. The corners will not have unsightly gaps or spaces.

A diagonal cut is made at each corner of a window. About $1/2$ inch is allowed around each opening. The diagonal cut forms a flap over the molding. This allows the opening to be cut for an exact fit. See Fig. 5-30F.

6
CHAPTER

Floor Preparation and Finish

THE FLOOR IS FINISHED LAST. THIS REALLY makes sense if you think about it. After all, people will be working in the building. They will be using paint, varnish, stain, plaster, and many other things. All these could damage a finished floor if they spilled on it.

Also, workers will be often entering and leaving the building. Mud, dirt, dust, and trash will be tracked into the building. It is very difficult to paint or plaster without spilling. A freshly varnished floor or a newly carpeted one could be easily ruined. So the floor is finished last to avoid damage to the finish floor.

Special methods are sometimes needed to lay flooring over concrete. Figure 6-1 shows some steps in this. Also, this was discussed in an earlier chapter.

For all types of floors, the first step is to clean them. See Fig. 6-2. The surface is scraped to remove all plaster, mud, and other lumps. Then the floor is swept with a broom.

Laying Wooden Flooring

Wooden flooring most often comes in three shapes. The first is called *strip* flooring (Fig. 6-3). The second is called *plank* flooring, as in Fig. 6-4. The third type is *block* or *parquet* flooring (Fig. 6-5). The blocks may be solid, as in Fig. 6-6A. Here the grain runs in one direction, and the blocks are cut with tongues and grooves. The type shown in Fig. 6-6B may be straight-sided. They are made of strips with the grain running in the same direction. Such blocks may have a spline at the bottom or a bottom layer.

The type of block shown in Fig. 6-6C is made of several smaller pieces glued to other layers. The bottom layer is usually waterproof.

Most flooring has cut tongue-and-groove joints. Hidden nailing methods are used so that the nails do not show when the floor is done. Strip and plank flooring will also have an undercut area on the bottom. This undercut is shown in Fig. 6-3. The undercut or *hollow* helps provide a stable surface for the flooring. Small bumps will not make the piece shift. Strip flooring has narrow, even widths with tongue-and-groove joints on the ends. The strips may be random end-matched. Plank flooring comes in both random widths and lengths. It may be drilled and pegged at the ends. However, today most plank flooring has fake pegs that are applied at the factory. The planks are then nailed in much the same manner as strip flooring. Block flooring may be either nailed or glued, but glue is used most often.

Most flooring is made from oak. Several grades and sizes are available. However, the width or size is largely determined by the type of flooring.

(A)

(B)

(C)

Fig. 6-1 *Laying wood floors on concrete. (A) Tar or asphalt is poured on. It acts as a vapor barrier. (B) Wood strips are positioned on the asphalt. (C) Flooring is nailed to the strips.* (National Oak Flooring Manufacturers Association)

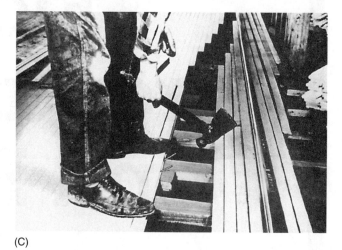

Fig. 6-2 *Subflooring must be scraped and cleaned.* (National Oak Flooring Manufacturers Association)

Carpeting can cost much more than a finished wood floor. Also, carpeting may last only a few years. As a rule, wood floors last for the life of the building. It is a good idea to have hardwood floors underneath carpet. Then the carpet can be removed without greatly reducing the resale value of the building.

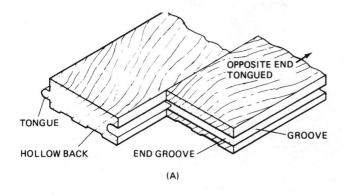

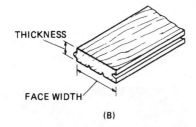

Fig. 6-3 *Strip flooring. (Forest Products Laboratory)*

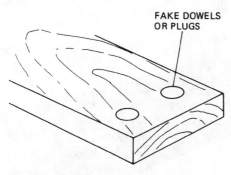

Fig. 6-4 *Plank flooring.*

Fig. 6-5 *A parquet floor of treated wood blocks. (Perma Grain Products)*

Preparation for Laying Flooring

Manufacturers recommend that flooring be placed inside for several days before it is laid. The bundles should be opened and the pieces scattered around the room. This lets the wood reach a moisture content similar to that of the room. This will help stabilize the flooring. If the flooring is stabilized, the expansion and contraction will be even.

Check the subfloor for appearance and evenness. The subfloor should be cleaned and scraped of all deposits and swept clean. Nails or nail heads should be removed. All uneven features should be planed or sanded smooth. See Fig. 6-2.

A vapor barrier should be laid over the subfloor. It can be made of either builder's felt or plastic film. Seams should be overlapped 2 to 4 inches. Then chalk lines should be snapped to show the centers of the floor joists. See Fig. 6-7.

Installing wooden strip flooring Wooden strip flooring should be applied perpendicular to the floor joists. See Fig. 6-8. The first strip is laid with the grooved edge next to the wall. At least ½ inch is left between the wall and the flooring. This space controls expansion. Wooden flooring will expand and contract. The space next to the wall keeps the floor from buckling or warping. Warps and buckles can cause air gaps beneath the floor. They also ruin the looks of the floor. The space next to the wall will be covered later by molding and trim.

The first row of strips is nailed using one method. The following rows are nailed differently. In the first row, nails are driven into the face. See Figs. 6-9 and 6-10. Later, the nail is set into the wood and covered.

Hardwood flooring will split easily. Most splits occur when a nail is driven close to the end of a board. To prevent this, drill the nail hole first. It should be slightly smaller than the nail.

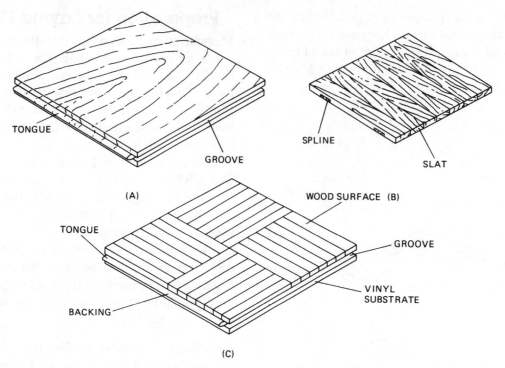

Fig. 6-6 *Wood block (parquet) flooring. (A) Solid.* (Forest Products Laboratory) *(B) Splined* (Forest Products Laboratory) *(C) Substrated.*

Fig. 6-7 *Vapor barriers are needed over board subfloors.* (National Oak Flooring Manufacturers Association)

The next strip is laid in place as shown in Fig. 6-11. The nail is driven blind at a 45° or 50° angle as shown. Note that the nail is driven into the top corner of the tongue. The second strip should fit firmly against the first layer. Sometimes, force must be used to make it fit firmly. A scrap piece of wood is placed over the tongue as shown in Fig. 6-12. The ends of the second layer should not match the ends of the first layer. The ends should be staggered for better strength and appearance. The end joints of one layer should be at least 6 inches from the ends on the previous layer. A nail set should be used to set the nail in place. Either the vertical position or the position shown in Fig. 6-12 may be used. Be careful not to damage the edges of the boards with the hammer.

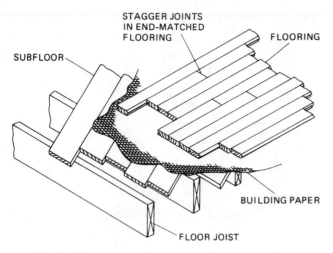

Fig. 6-8 *Strip flooring is laid perpendicular to the joists.* (Forest Products Laboratory)

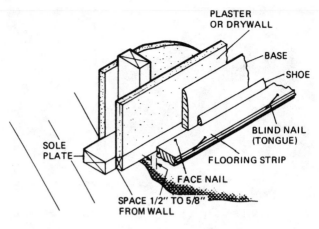

Fig. 6-10 *In the first strip, nails are driven into the face.* (Forest Products Laboratory)

Fig. 6-9 *Nailing the first strip.* (National Oak Flooring Manufacturers Association)

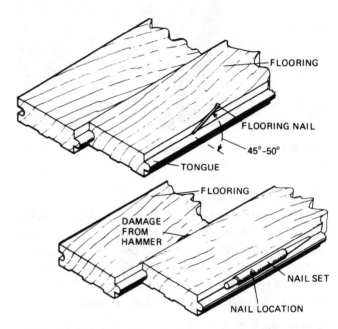

Fig. 6-11 *Nailing strips after the first.* (Forest Products Laboratory)

The same amount of space (½ inch) is left at the ends of the rows. End pieces may be driven into place with a wedge. Pieces cut from the ends can become the first piece on the next row. This saves material and helps stagger the joints. Figures 6-13 and 6-14 show other steps.

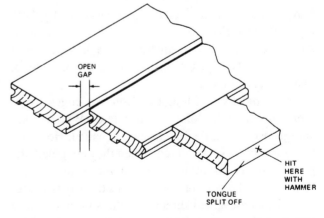

Fig. 6-12 *A scrap block is used to make flooring fit firmly. This prevents damage to either tongues or grooves.*

Fig. 6-13 *Boards are laid in position and nailed in place.* (National Oak Flooring Manufacturers Association)

Fig. 6-14 *The last piece is positioned with a pry bar. Nails are driven in the face.* (National Oak Flooring Manufacturers Association)

Wooden plank flooring Wooden plank flooring is installed in much the same manner. Today, both types are generally made with tongue-and-groove joints. The joints are on both the edges and ends. The same allowance of ½ inch is made between plank flooring and the walls. The same general nailing procedures are used.

Wood block floors There are two types of wood block floors. The first type is like a wide piece of board. The second type is called *parquet*. Parquet flooring is made from small strips arranged in patterns. Parquet must be laminated to a base piece. Today, both types of block floors are often laminated. When laminated, they are plywood squares with a thick veneer of flooring on the top. Usually, three layers of material make up each block.

Most blocks have tongue-and-groove edges. Common sizes are 3-, 5-, and 7-inch squares. With tongue-and-groove joints, the blocks can be nailed. However, today most blocks are glued. When they are glued, the preparation is different from that for plank or strip flooring. No layer of builder's felt or tar paper is used. The blocks are glued directly to the subfloor or to a base floor. As a rule, plywood or chipboard is used for a base under the block flooring.

Blocks may be laid from a wall or from a center point. Some patterns are centered in a room. Then blocks should be laid from the center toward the edges. To lay blocks from the center, first the center point is found. Then a block is centered over this point. The outline of the block is drawn on the floor. Then a chalk line is snapped for each course of blocks. Care is taken that the lines are parallel to the walls.

Sometimes blocks are laid on the floor, proceeding from a wall. The chalk lines should be snapped for each course from the base wall.

Blocks are often glued directly to concrete floors. These floors must be properly made. Moisture barriers and proper drainage are essential. When there is any doubt, a layer process is used. First, a layer of mastic cement is applied. Then a vapor barrier of plastic or felt is applied over the mastic. Then a second layer of mastic is applied over the moisture barrier. The blocks are then laid over the mastic surface.

Wooden blocks may also be glued over diagonal wood subflooring. The same layer process as above is used.

FINISHING FLOORS

In most cases, floors are finished after walls and trim. When resilient flooring is applied, no finish step is needed. However, floors of this type should be cleaned carefully.

Finishing Wood Floors

The first step in finishing wood floors is sanding. A special sanding machine is used, such as the one in Fig. 6-15. When floors are rough, they are sanded twice. For oak flooring, it is a good idea to use a sealer coat next. The sealer coat has small particles in it. These help fill the open pores in the oak. One of two types of sealer or filler should be used. The first type of sealer is a mixture of small particles and oil. This is rubbed into the floor and allowed to sit a few minutes.

Fig. 6-15 *After the floor is laid, it is sanded smooth using a special sanding machine.* (National Oak Flooring Manufacturers Association)

Then a rotary sander or a polisher is used to wipe off the filler. The filler on the surface is wiped off. The filler material in the wood pores is left.

Floors may be stained a darker color. The stain should be applied before any type of varnish or lacquer is used. It is a good idea to stain the molding at the same time. The stain is applied directly after the sanding. If a particle-oil type of filler is used, the stain should be applied after the filler.

The second filling method uses a special filler varnish or lacquer. This also contains small particles which help to fill the pores. It is brushed or sprayed onto the floor and allowed to dry. The floor is then buffed lightly with an abrasive pad.

After filling, the floor should be varnished. A hard, durable varnish is best. Other coatings are generally not satisfactory. Floors need durable finishes to avoid showing early signs of wear.

Molding can be applied after the floors are completely finished. The base shoe and baseboard are

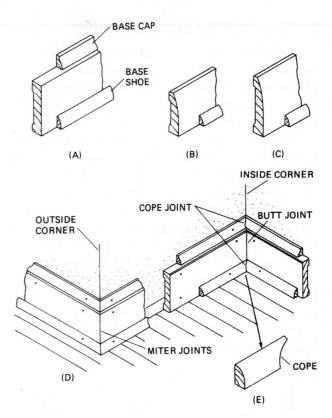

Fig. 6-16 *Base molding* (Forest Products Laboratory): *(A) Square-edged base; (B) narrow ranch base; (C) wide ranch base; (D) installation; (E) a cope joint* (Forest Products Laboratory).

installed in Fig. 6-16. Many workers also finish the molding when the floor is finished.

Base Flooring for Carpet

Often two floor layers are used but neither is the "finished" floor surface. The first layer is the subfloor, and the next layer is a base floor. Both base and subflooring may be made from underlayment. Underlayment may be a special grade of plywood or chipboard. In many cases, nailing patterns are printed on the top side of the underlayment.

It is a good idea to bring underlayment into the room to be floored and allow it to sit for several days, exposed to the air. This allows it to reach the same moisture content as the rest of the building components.

Base flooring is used when added strength and thickness are required. It is also used to separate resilient flooring or other flooring materials from concrete or other types of floors. It provides a smooth, even base for carpet. It is much cheaper to use a base floor than a hardwood floor under carpets. The costs of nailing small strips and of sanding are saved.

INSTALLING CARPET

Carpet is becoming more popular for many reasons. It makes the floor a more resilient and softer place for people to stand. Also, carpeted floors are warmer in winter. Carpeting also helps reduce noise, particularly in multistory buildings.

To install carpet, several factors must be considered. First, most carpet is installed with a pad beneath it. When carpet is installed over a concrete floor, a plastic film should also be laid. This acts as an additional moisture seal. The carpet padding may then be laid over the film.

The first step in laying carpet is to attach special carpet strips. These are nailed to the floor. They are laid around the walls of a room to be carpeted. These carpet strips are narrow, thin pieces of wood with long tacks driven through them. They are nailed to the floor approximately ¼ inch from the wall. See Fig. 6-17. The carpet padding is then unrolled and cut. The padding extends only to the strips.

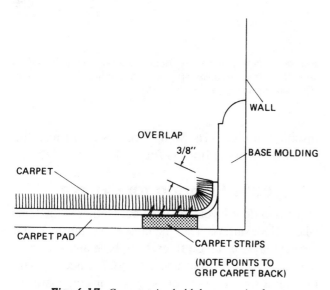

Fig. 6-17 *Carpet strips hold the carpet in place.*

Next, the carpet is unrolled. If the carpet is large enough, it may be cut to exactly fit. However, carpet should be cut about 1 inch smaller than the room size.

The carpet is wedged between the carpet strip and one wall. A carpet wedge is used as in Fig. 6-18. The carpet is then smoothed toward the opposite wall. All wrinkles and gaps are smoothed and removed. Next, the carpet is stretched to the opposite wall. The person installing the carpet will generally walk around the edges, pressing the carpet into the tacks of the carpet strips. This is done after each side is wedged. After the ends are wedged, the first side is attached. The same process is repeated. After the first side is wedged, the opposite side is attached.

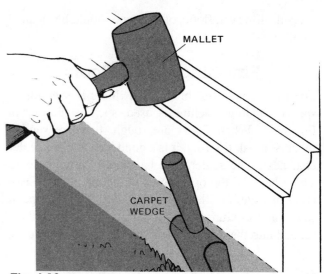

Fig. 6-18 *Carpet edges are wedged into place. Base shoe molding may then be added.*

To seam a carpet, special tape is used. This tape is a wide strip of durable cloth. It has an adhesive on its upper surface. It is rolled out over the area or edge to be joined. Half of the tape is placed underneath the carpet already in place. The carpet is then firmly pressed onto the adhesive. Some adhesive tapes use special heating tools for best adhesion. The second piece of carpet is carefully butted next to the first. Be sure that no great pressure is used to force the two edges together. No gaps should be wider than ¹/₁₆ inch. The edges should not be jammed forcefully together, either. If edges are jammed together, lumps will occur. If wide gaps are left, holes will occur. However, the nap or shag of the carpet will cover most small irregularities.

To cut the carpet for a joint, first unroll the carpet. Carefully size and trim the carpet edges as straight as possible. The joint will not be even if the edges are ragged. Various tools may be used. A heavy knife or a pair of snips may be used effectively.

Metal end strips are used where carpet ends over linoleum or tile. The open end strips are nailed to the floor. The carpet is stretched into place over the points on the strip. Then the metal strip is closed. A board is laid over the strip. See Fig. 6-19. The board is struck sharply with a hammer to close the strip. Do not strike the metal strip directly with the hammer. Doing so will leave unsightly hammer marks on the metal.

RESILIENT FLOORING

Resilient flooring is made from chemicals rather than from wood products. Resilient flooring includes compositions such as linoleum and asphalt tile. It may be laid directly over concrete or over base flooring. Resilient flooring comes in both sheets and square tiles.

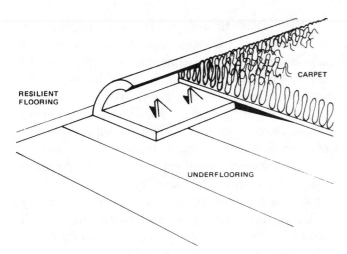

Fig. 6-19 *A metal binder bar protects and hides carpet edges where carpet ends over linoleum or tile.*

Frequently, resilient flooring sheets will be called *linoleum carpets*.

Installing Resilient Flooring Sheets

The first step is to determine the size of the floor to be covered. It is a good idea to sketch the shape on a piece of paper. Careful measurements are made on the floor to be covered. Corners, cabinet bases, and other features of the floor are included. It is a good idea to take a series of measurements. They are made on each wall every 2 or 3 feet. This is needed because most rooms are not square. Thus, measurements will vary slightly from place to place. The measurements should be marked on the paper.

Next, the floor is cleaned. Loose debris is removed. A scraper is used to remove plaster, paint, or other materials. Then the area is swept. If necessary, a damp mop is used to clean the area. Neither flooring nor cement will stick to areas that are dirty. Next, the surface is checked for holes, pits, nail heads, or obstructions. Holes larger than the diameter of a nail are patched or filled. Nails or obstructions are removed.

Most rooms will be wider than the roll of linoleum carpet. If not, the outline of the floor may be transferred directly to the flooring sheet. The sheet may then be cut to shape. The shaped flooring sheet is brought into the room. It is positioned and unrolled. A check is made for the proper fit and shape. Any adjustments or corrections are made. Then the sheet is rolled up approximately halfway. The mastic cement is spread evenly (about $3/32$ inch thick) with a toothed trowel. The unrolled portion is rolled back into place over the cemented area. Then the other end is rolled up to expose the bare floors. Next, the mastic is spread over the remaining part of the floor. The flooring is rolled back

into place over the cement. The sheet is smoothed from the center toward the edge.

However, most rooms are wider than the sheets of flooring. This means that two or more pieces must be joined. It is best to use a factory edge for the joint line. Select a line along the longest dimension of the room as shown in Fig. 6-20. On the base floor, measure equal distances from a reference wall as shown. These are the same width as the sheet. Then snap a chalk line for this line. Often more than one joint or seam must be used.

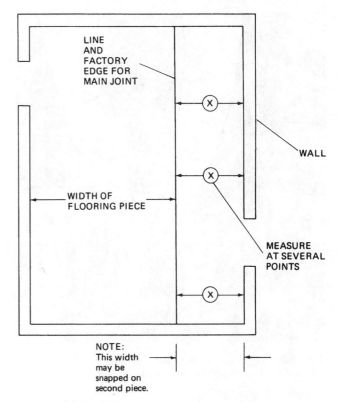

Fig. 6-20 *Floor layout for resilient flooring.*

The same measuring process is used. Measurements are taken from the edge of the first sheet. However, for smaller pieces, a different process is used. The center of the last line is found. A line is snapped at right angles. This shows the pattern for the pieces of flooring that must be cut. The second line should run the entire width of the room. A carpenter's square may be used to check the squareness of the lines.

The two chalk lines are now the reference lines. These are used for measuring the flooring and the room area. Measurements to the cabinets or other features are made from these lines. Take several measurements from the upper layout. Walls are seldom square or straight. Frequent measurements will help catch these irregularities. The fit will be more accurate and better.

Next, in a different area unroll the first sheet of flooring to be used. Find the corresponding wall and reference line. The factory edge is aligned on the first line. The dimensions are marked on the resilient flooring. The necessary marks show the floor outline on the flooring sheet. A straight blade of linoleum knife is used to make these cuts. For best results, a guide is used. A heavy metal straightedge is used for a guide when cutting. A check is made to be sure that there is nothing underneath the flooring. Anything beneath it could be damaged by the knife used to cut the flooring.

Fig. 6-21 *Edges are overlapped during cutting. This way seams will match even if the cut is not perfectly straight.* (Armstrong Cork)

Next, the cut flooring piece is carried into the area. It is unrolled over the area, and the fit is checked. Any adjustments necessary are made at this point. Next, the material is rolled toward the center of the room. The area is spread with mastic. The flooring is rolled back into place over the cement. The sheet is smoothed as before. Do not force the flooring material under offsets or cabinets. Make sure that the proper cuts are made.

The adhesive should not be allowed to dry more than a few minutes. No more than 10 or 15 minutes should pass before the material is placed. A heavy roller is recommended to smooth the flooring. It should be smoothed from the center toward the edges. This removes air pockets and bubbles.

Where seams are made, a special procedure is recommended. Unroll the two pieces of flooring in the same preparation area. The two edges to be joined are slightly overlapped. See Fig. 6-21. Next, the heavy metal guide is laid over the doubled layer. Then both layers are cut with a single motion. The straightedge is used as a guide to make the straightest possible cut. Edges cut this way will match, even if they are not perfectly straight or square. Figure 6-22 shows how edges are trimmed.

To Install Resilient Block Flooring

Resilient block flooring is often called *tile.* This term includes floor tile, asphalt tile, linoleum tile, and others. As a rule, the procedure for all these is the same. The sizes of the tiles range from 6 to 12 inches square. To lay tile, the center of each of the end walls is found. See Fig. 6-23. A chalk line is tied to each of these points. Lines are then snapped down the middle of the

Fig. 6-22 *Trimming the edges. A metal straightedge or carpenter's square is used to guide a utility knife.* (Armstrong Cork)

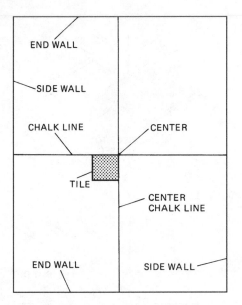

Fig. 6-23 *Find the intersecting midpoints.* (Armstrong Cork)

floor. Next, the center of the first line is located. A square or another tile is used as a square guide. A second line is snapped square to the first line.

Next, a row of tiles is laid along the perpendicular chalk line as shown. They are not cemented. Then the distance between the wall and the last tile is measured. If the space is less than one-half the width of a tile, a new line is snapped. It is placed one-half the width of a tile away from the centerline. See Fig. 6-24. A second line is snapped one-half the width of a tile from the perpendicular line. The first tile is then aligned on the second snapped line. The tile can now be cemented. The first tile becomes the center tile of the room.

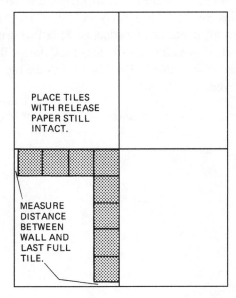

Fig. 6-24 *Lay tiles in place without cement to check the spacing.*
(Armstrong Cork)

Another method is used if the distance at the side walls is greater than one-half the width of a tile. Then the tile is laid along the first line. This way, no single tile is the center.

The first two courses are laid as a guide. Then the mastic is spread over one-quarter of the room area. See Fig. 6-25. Lay the tiles at the center first. The first tile should be laid to follow the snapped chalk lines. Tiles should not be slid into place. Instead, they are pressed firmly into position as they are installed. It is best to hold them in the air slightly. Then the edges are touched together and pressed down.

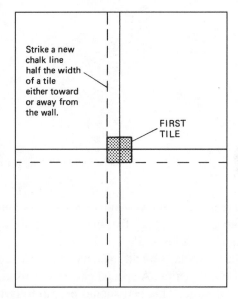

Fig. 6-25 *The next chalk line is used as a guide for laying tile. This method places one tile in the center.*

The first quarter of the room is covered. Only the area around the wall is open. Here, tiles must be cut to fit.

Cutting the tiles   A loose tile is placed exactly on top of the last tile in the row. See Fig. 6-26A. Then a third tile is laid on top of this stack. It is moved over until it touches the wall. The edge of the top of the tile becomes the guide. See Fig. 6-26B. Then the middle tile is cut along the pencil line as shown. This tile will have the proper spacing.

A pattern is made to fit tile around pipes or other shapes. The pattern is made in the proper shape from paper. This shape is traced on the tile. The tile is then cut to shape.

LAYING CERAMIC TILE

Several types of ceramic tile are commonly used: ceramic tile, quarry tile, and brick. Tiles are desirable where water will be present, such as in bathrooms and

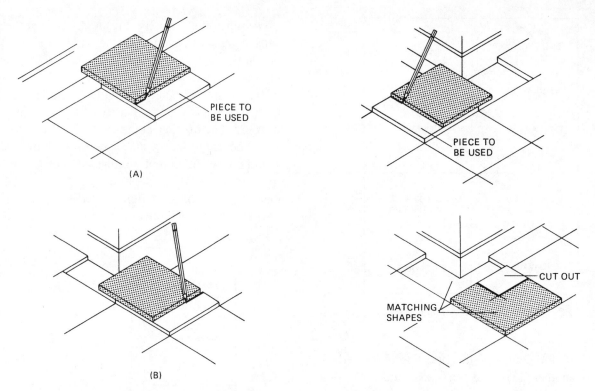

PIECE TO
BE USED

(A)

PIECE TO
BE USED

(B)

CUT OUT

MATCHING
SHAPES

Fig. 6-26 *Cutting and trimming tile to fit. (A) Marking tile for edges. (B) Marking tile for corners.*

wash areas. However, ceramic tiles also make pleasing entry areas and lobbies. Two methods are used for installing ceramic floor tile.

The first, and more difficult, method uses a cement-plaster combination. A special concrete is used as the bed for the tile. This bed should be carefully mixed, poured, leveled, and troweled smooth. It should be allowed to sit for a few minutes. Just before it hardens, it is still slightly plastic. Then the tiles are embedded in place. The tiles should be thoroughly soaked in water if this method is used. They should be taken out of the water one at a time and allowed to drain slightly. The tiles are then pressed into place in the cement base. All the tiles are installed. Then special grout is pressed into the cracks between the tiles. This completely fills

the joints between the tiles. The grouted joints should be cleaned and tooled within a very few hours after installation.

Adhesives are most widely used today. Adhesives are much easier to use. The adhesives are spread evenly on the tiles. This is much like laying resilient tile. The adhesive should be one that is recommended for use with ceramic tiles. After the adhesive has been spread, the tiles are placed. Then a common grout is forced between the tiles. It is wiped and allowed to dry.

Tiles are available as large individual pieces. They are also available as assortments. Small tiles may be preattached to a cloth mesh. This mesh keeps the tiles arranged in the pattern. This also makes the tile easy to space evenly.

7
CHAPTER

Kitchens and Cabinetry

Kitchen Planning

The kitchen is the focal point for family life in most homes. See Fig. 7-1. Some sources estimate that family members spend up to 50% of their time in the kitchen. For this reason, this is generally the most used and the most remodeled room of a home. Remodeling calls for a carpenter to be able to rearrange according to the desires of the owners.

The kitchen is a physically complex area in that it sustains heavy traffic flow from people using it and passing through it. It contains hot and cold water sources, drains, plumbing, and electrical outlets and fixtures; and it is exposed to high moisture, splashing water and other liquids, items that are intensely hot, and items that are very cold. In addition, activities involve sharp instruments, and considerable forces may be exerted in blending, rolling, pressing, cutting, and shaping. Other activities include mixing, washing, transferring, and storing.

Four major functions must be considered in kitchen planning: cooking, storing, eating, and entertaining. While most people consider cleaning up as part of cooking, it should be considered separately. The kitchen may also be the base of operations for several other functions done by one or more family members: studying, using a computer, home office work, sewing, and laundry. A larger kitchen may sometimes include home entertainment equipment or be used as an area for children.

The work area in a kitchen must be considered in laying out the cabinets, refrigerator, range, and sink.

The work triangle is very important since that is where most of the food preparation is done. See Fig. 7-2.

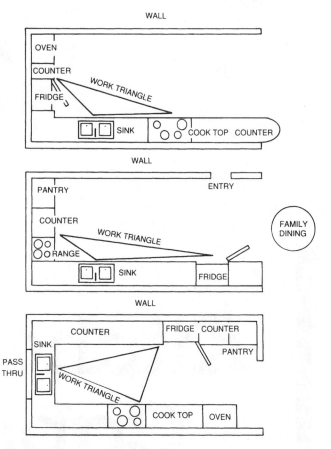

Fig. 7-2 *Location of the work triangle for three different kitchen arrangements.*

Fig. 7-1 *Efficient arrangement of the kitchen cooking area.* (Jenn-Air)

Cabinets may be finished in a variety of ways. They should be coordinated with the kitchen fixtures. Wooden cabinets with white plastic laminate covers can be made or purchased. These lend a European style that contrasts the white and the warm wood tones. See Fig. 7-3 for the rough layout before the tile is placed on the countertop. Figure 7-4 shows how the finished countertop looks. As you can see from these illustrations, the cabinets can be made on the job or purchased and the finishing touches made by the carpenter on the job.

Countertops   Countertops may be built as part of the cabinet and then covered with almost any counter material, or they may be purchased as slabs cut to the right length. Openings in either type may be cut out.

Fig. 7-3 *Plywood base installed for the application of the countertop tile. (American Olean Tile)*

These slabs may have a splashboard molded in as part of the slab, or the splashboard may be added later. Some building codes require all sinks and basins to have splashboards; others do not. See Fig. 7-5.

Kitchen light   Kitchens need lots of light to be most useful and enjoyable. It is a good idea to have a good general light source combined with several additional lights for specific areas. Areas in which lights are necessary include the sink, cooktop, under cabinets, and in areas where recipes or references are kept. Figure 7-6 is a good example of a kitchen with a skylight and plenty of cooking and work areas.

Kitchen safety   Unfortunately, kitchens are the scene of many accidents. Sharp objects, wet floors, boiling pots, and open flames are all potential hazards.

Cooktops should never be installed under windows because most people hang curtains, shades, or blinds in them. These can catch fire from the flames or heating elements. In addition, people lean over to look out a window or to open or shut it. This puts them directly over the cooktop where they could be burned.

Cooktops should have 12 to 15 inches of counter around all sides. This keeps the handles of the pots from extending beyond the edges where they can be bumped by passing adults or grabbed by curious children. Either situation can cause bad spills and nasty burns.

Using revolving shelves or tiered and layered swing-out shelves can reduce falls caused by trying to reach the back of top shelves.

Heavy objects should be stored at or near the level at which they are used. Above all, they should not be stored high overhead or above a work area.

Fig. 7-4 *Finished countertop. (American Olean Tile)*

Fig. 7-5 *Note the splashboards on these countertops.*

Fig. 7-6 *Ceramic tile is used on the floors and the walls as well as the countertop to provide a durable kitchen with plenty of light.* (American Olean Tile)

CABINETS AND MILLWORK

Millwork is a term used for materials made at a special factory, or mill. It includes both single pieces of trim and big assemblies. Interior trim, doors, kitchen cabinets, fireplace mantels, china cabinets, and other units are all millwork. Most of these items are sent to the building ready to install. So the carpenter must know how to install units.

However, not all cabinets and trim items are made at a mill. The carpenter must know how to both construct and install special units. This is called *custom* work. Custom units are usually made with a combination of standard-dimension lumber and molding or trim pieces. Also, many items that are considered millwork do not require a highly finished appearance. For example, shelves in closets are considered millwork. But they generally do not require a high degree of finish. On the other hand, cabinets in kitchens or bathrooms do require better work.

Various types of wood are used for making trim and millwork. If the millwork is to be painted, pine or other soft woods are used most often. However, if the natural wood finish is applied, hardwood species are generally preferred. The most common woods include birch, mahogany, and ash. Other woods such as poplar and boxwood are also used. These woods need little filling and can be stained for a variety of finishes.

Installing Ready-Built Cabinets

Ready-built cabinets are used most in the kitchens and bathrooms. They may be made of metal, wood, or wood products. The carpenter should remember that overhead cabinets may be used to hold heavy dishes and appliances. Therefore, they must be solidly attached. Counters and lower cabinets must be strong enough to support heavy weights. However, they need not be fastened to the wall as rigidly as the upper units. Ready-builts are obtained in widths from 12 to 48 inches. The increments are 3 inches. Thus cabinets of 12, 15, 18, and 21 inches, and so forth, are standard. They may be easily obtained. Figure 7-7 shows some typical kitchen cabinet dimensions.

Ordering cabinets takes careful planning. There are many factors for the builder or carpenter to consider. The finish, the size and shape of the kitchen, and the dimensions of built-in appliances must be considered.

For example, special dishwashing units and ovens are often built in. The cabinets ordered should be wide enough for these to be installed. Also, appliances may be installed in many ways. Manufacturer's data for both the cabinet and the appliance should be checked. Special framing may be built around the appliance. The frame can then be covered with plastic laminate materials. This provides surfaces that are resistant to heat, moisture, and scratching or marring. See Fig. 7-8A and B.

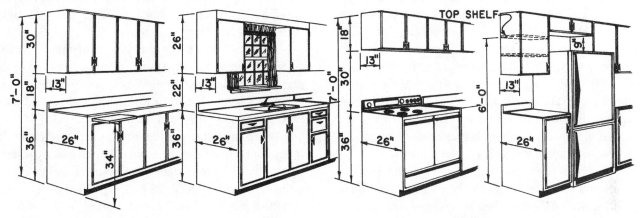

Fig. 7-7 *Typical cabinet dimensions.*

(A)   (B)

Fig. 7-8 *(A) Appliance is built-in to extend. (B) Appliance is built-in flush.*

Sinks, dishwashers, ranges, and refrigerators should be carefully located in a kitchen. These locations are important in planning the installation of cabinets. Plumbing and electrical connections must also be considered. Natural and artificial lighting can be combined in these areas.

Cabinets and wall units should have the same standard height and depth. It would be poor planning to have wall cabinets with different widths in the same area.

Five basic layouts are commonly used in the design of a kitchen. These include the sidewall type, as shown in Fig. 7-9A. This type is recommended for small kitchens. All elements are located along one wall.

The next type is the parallel or pullman kitchen. See Fig. 7-9B. This is used for narrow kitchens and can be quite efficient. The arrangement of the sink, refrigerator, and range is critical for efficiency. This type of kitchen is not recommended for large homes or families. Movement in the kitchen is restricted, but it is efficient.

The third type is the L shape. See Fig 7-9C. Usually the sink and the range are on the short legs. The refrigerator is located on the other leg. This type of arrangement allows for an eating space on the open end.

The U arrangement usually has a sink at the bottom of the U. The range and refrigerator should be located on opposite sides for best efficiency. See Fig. 7-9D.

The final type is the island kitchen. This type is becoming more and more popular. It promotes better utilization and has better appearance. This arrangement makes a wide kitchen more efficient. The island is the central work area. From it, the appliances and other work areas are within easy reach. See Fig. 7-9E.

Screws should be used to hang cabinet units. The screws should be at least No. 10 three-inch screws. The screws reach through the hanging strips of the cabinet. They should penetrate into the studs of the wall frame. Toggle bolts could be used when studs are inaccessible. However, the walls must be made of rigid materials rather than plasterboard.

To install wall units, one corner of the cabinet is fastened. The mounting screw is driven firm; the other end is then plumbed level. While someone holds the cabinet, a screw is driven through at the second end. Next, screws are driven at each stud interval.

When installing counter units, care must be taken in leveling. Floors and walls are not often exactly square or plumb. Therefore, care must be taken to install the unit plumb and level. What happens when a unit is not installed plumb and level? The doors will not open properly, and the shelves will stick and bind.

Shims and blocking should be used to level the cabinets. Shingles or planed blocks are inserted beneath the cabinet bases.

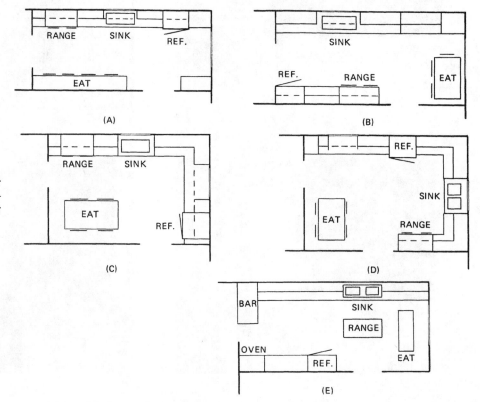

Fig. 7-9 *Basic kitchen layouts: (A) Sidewall; (B) parallel (pullman); (C) L shape; (D) U shape; (E) island.* (Forest Products Laboratory)

The base unit and the wall unit are installed first. Then the countertop is placed. Countertops may be supplied as prefabricated units. These include dashboard and laminated tops. After the countertops are applied, they should be protected. Cardboard or packing should be put over the top. It can be taped in place with masking tape and removed after the building is completed. Sometimes the cabinets are hung but not finished. The countertop is installed after the cabinets are properly finished.

Ready-built units may be purchased three ways. First, they may be purchased assembled and prefinished. Counters are usually not attached. The carpenter must install them and protect the finished surfaces.

Second, cabinets may be purchased assembled but unfinished. These are sometimes called *in-the-white* and are very common. They allow all the woodwork and interior to be finished in the same style. Such cabinets may be made of a variety of woods. Birch and ash are among the more common hardwoods used. The top of the counter is usually provided but not attached. Also, the plastic laminate for the countertop is not provided. A contractor or carpenter must purchase and apply this separately.

Third, the cabinets are purchased unassembled. The parts are precut and sanded. However, the unit is shipped in pieces. These are put together and finished on the job.

Cabinets are sometimes a combination of special and ready-built units. Many builders use special crews who do only this type of finish work. The combination of counter types gives a specially built look with the least cost.

Making Custom Cabinets

Custom cabinets are special cabinets that are made on the job. Many carpenters refer to these and any type of millwork as *built-ins*. See Fig. 7-10. These jobs include building cabinets, shelves, bookcases, china closets, special counters, and other items.

A general sequence can be used for building cabinets. The base is constructed first. Then a frame is made. Drawers are built and fitted next. Finally the top is built and laminated.

Pattern layout Before beginning the cabinet, a layout of the cabinet should be made. This may be done on plywood or cardboard. However, the layout should be done to full size if possible. The layout should show sizes and construction methods involved. Figure 7-11 shows a typical layout.

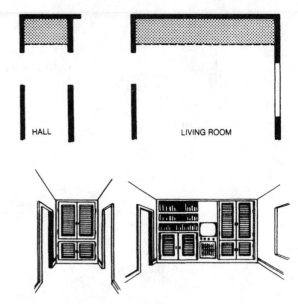

Fig. 7-10 *Typical built-ins.*

Custom cabinets can be made in either of two ways. The first involves cutting the parts and assembling them *in place*. Each piece is cut and attached to the next piece. When the last piece is done, the cabinet is in place. The cabinet is not moved or positioned.

For the second procedure, all parts are cut first. The cabinet unit is assembled in a convenient place. Then it is moved into position. The cabinet is leveled and plumbed and then attached.

Several steps are common to cabinet making. A bottom frame is covered with end and bottom panels. Then partitions are built and the back top strip is added. Facing strips are added to brace the front. Drawer guides and drawers are next. Shelves and doors complete the base.

Making the base The base for the cabinets is made first. Either 2- × 4- or 1- × 4-inch boards may be used for this. No special joints are used. See Fig. 7-12. Then the end panel is cut and nailed to the base. The toeboard, or front, of the base covers the ends. The end panel covers the end of the toeboard. A temporary brace is nailed across the tops of the end panels. This braces the end panels at the correct spacing and angles.

Bottom panels are cut next. The bottom panels serve as a floor over the base. The partition panels are placed next. They should be notched on the back top. This allows them to be positioned with the back top strip.

The back top strip is nailed between the end panels as shown. The temporary brace may then be removed. The partition panels are placed into position and nailed to the top strip. For a cabinet built in position, the

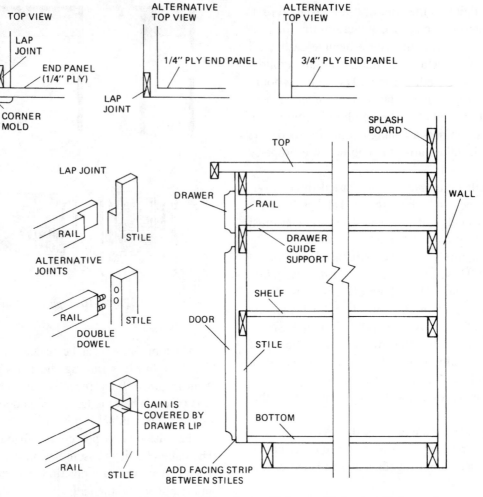

Fig. 7-11 *A cabinet layout. Dimensions and joints would also be added.*

partitions are toenailed to the back strip. The process is different if the cabinet is not built in place. The nails are driven from the back to the edges of the partitions. The temporary braces may be removed.

Cutting facing strips Facing strips give the unit a finished appearance. They cover the edges of the panels, which are usually made of plywood. This gives a better and more pleasing appearance. The facing strips also brace and support the cabinet and support the drawers and doors. Because they are supports, special notches and grooves are used to join them.

The vertical facing strip is called a *stile*. The horizontal piece is called the *rail*. They are notched and joined as shown in Fig. 7-13. Note that two types of rail joints are used. The flat or horizontal type uses a notched joint. The vertical rail type uses a notched lap joint in both the rail and the stile.

As a rule, stiles are nailed to the end and partition panels first. The rails are then inserted from the rear. Glue may be used, but nails are preferred. Nail holes

are drilled for nails near the board ends. The hole is drilled slightly smaller than the nail diameter. Finishing nails should be driven in place. Then the head is set below the surface. The hole is later filled.

Drawer guides After the stiles and rails, the drawer assemblies are made. The first element of the drawer assembly is the drawer guide. This is the portion into which the drawer is inserted. The drawer guides act as support for and guide the drawers as they move in and out. The drawer guide is usually made from two pieces of wood. It provides a groove for the side of the drawer. The side of the guide is made from the top piece of wood. It prevents the drawer from slipping slideways.

A strip of wood is nailed to the wall at the back of the cabinet. It becomes the back support for the drawer guide. The drawer guide is then made by cutting a bottom strip. This fits between the rail and the wall. The top strip of the guide is added next. See Fig. 7-13. It becomes the support for the drawer guide at the front of the cabinet. Glue and nails are used to assemble this

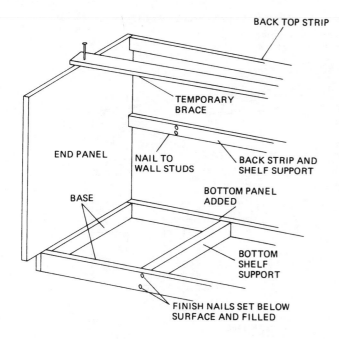

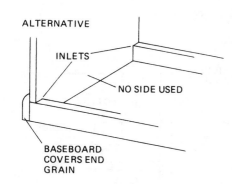

Fig. 7-12 The base is built first. Then end panels are attached to it.

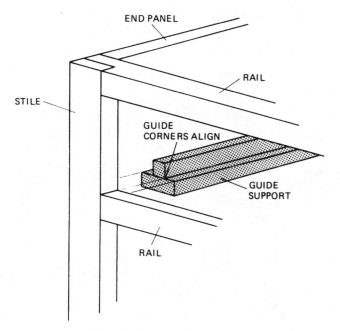

Fig. 7-13 Drawer guide detail.

unit. The glue is applied in a weaving strip. Finish nails are then driven on alternate sides about 6 inches apart to complete the assembly. Each drawer should have two guides. One is on each side of the drawer.

Three common types of drawer guides are made by carpenters: side guides, corner guides, and center guides. Special guiderails may also be purchased and installed. These include special wheels, end stops, and other types of hardware. A typical set of purchased drawer guides is shown in Fig. 7-14.

The corner guide has two boards that form a corner. The bottom edge of the drawer rests in this corner. See Fig. 7-15A. A side guide is a single piece of wood nailed to a cabinet frame. As in Fig. 7-15B, the single piece of wood fits into a groove cut on the side of the drawer. The guides serve both as a support and as guides.

For the center guide, the weight of the drawer is supported by the end rails. However, the drawer is kept in alignment by a runner and guide. See Fig. 7-15C. The carpenter may make the guides by nailing two strips of wood on the bottom of the drawer. The runner is a single piece of wood nailed to the rail.

In addition to the guides, a piece should be installed near the top. This keeps the drawer from falling as it is opened. This piece is called a *kicker.* See Fig. 7-16. Kickers may be installed over the guides. Or, a single kicker may be installed.

Drawer guides and drawer rails should be sanded smooth. Then they should be coated with sanding sealer. The sealer should be sanded lightly and a coat of wax applied. In some cases, the wood may be sanded and wax applied directly to the wood. However, in either case, the wax will make the drawer slide better.

Making a drawer   Most drawers made by carpenters are called *lip drawers.* This type of drawer has a lip around the front. See Fig. 7-17. The lip fits over the drawer opening and hides it. This gives a better appearance. It also lets the cabinet and opening be less accurate.

The other type of drawer is called a *flush drawer.* Flush drawers fit into the opening. For this reason, they must be made very carefully. If the drawer front is not accurate, it will not fit into the opening. Binding or large cracks will result. It is far easier to make a lip drawer.

Drawers may be made in several ways. As a rule, the procedure is to cut the right and left sides as in Fig. 7-18. The back ends of these have dado joints cut into them. Note that the back of the drawer rests on the bottom piece. The bottom is grooved into the sides and front piece as shown.

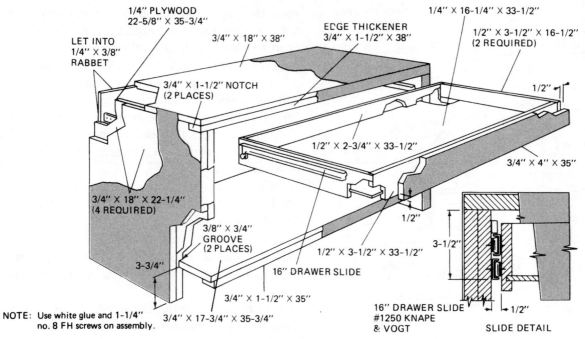

Fig. 7-14 *A built-in unit with factory-made drawer guides.* (Formica)

1/4" PLYWOOD
22-5/8" X 35-3/4"

LET INTO
1/4" X 3/8"
RABBET

3/4" X 18" X 38"

EDGE THICKENER
3/4" X 1-1/2" X 38"

1/4" X 16-1/4" X 33-1/2"

1/2" X 3-1/2" X 16-1/2"
(2 REQUIRED)

3/4" X 1-1/2" NOTCH
(2 PLACES)

1/2"

1/2" X 2-3/4" X 33-1/2"

3/4" X 4" X 35"

3/4" X 18" X 22-1/4"
(4 REQUIRED)

3/8" X 3/4"
GROOVE
(2 PLACES)

1/2"

3-3/4"

16" DRAWER SLIDE

1/2" X 3-1/2" X 33-1/2"

3-1/2"

NOTE: Use white glue and 1-1/4"
no. 8 FH screws on assembly.

3/4" X 1-1/2" X 35"

3/4" X 17-3/4" X 35-3/4"

16" DRAWER SLIDE
#1250 KNAPE
& VOGT

1/2"

SLIDE DETAIL

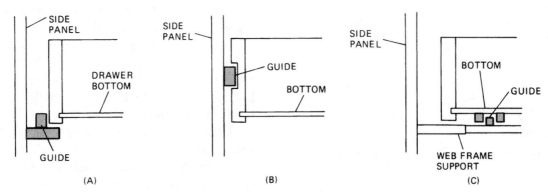

SIDE
PANEL

DRAWER
BOTTOM

GUIDE

(A)

SIDE
PANEL

GUIDE

BOTTOM

(B)

SIDE
PANEL

BOTTOM

GUIDE

WEB FRAME
SUPPORT

(C)

Fig. 7-15 *Carpenter-made drawer guides; (A) Corner guide; (B) side guide; (C) center guide.*

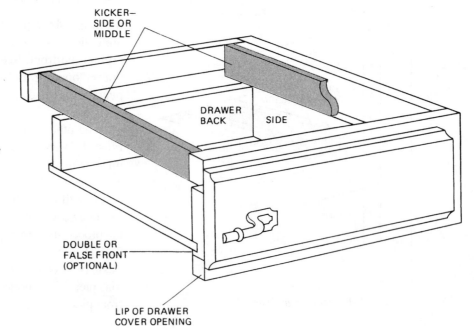

KICKER—
SIDE OR
MIDDLE

DRAWER
BACK

SIDE

Fig. 7-16 *Kickers keep draws from tilting out when opened.*

DOUBLE OR
FALSE FRONT
(OPTIONAL)

LIP OF DRAWER
COVER OPENING

The front of the drawer should be made carefully. It takes the greatest strain from opening and closing. The front should be fitted to the sides with a special joint. Several types of joint may be used. These are shown n Fig. 7-19.

The highest-quality work can feature a special joint called the *dovetail joint.* However, this joint is expensive to make and cut. As a rule, other types of dovetail joints are used. The dovetailed dado joint is often used.

Drawers are made after the cabinet frame has been assembled. Fronts of drawers may be of several shapes. The front may be paneled, as in Fig. 7-20A, or it can be molded, as in Fig. 7-20B. Both styles can be made thicker by adding extra boards.

The wood used for the drawer front is selected and cut to size. Joints are marked and cut for assembly. Next, the groove for the bottom is cut.

Then the stock is selected and cut for the sides. As a rule, the sides and backs are made from different wood than the front. This is to reduce cost. The drawer front may be made from expensive woods finished as required. However, the interiors are made from less ex-

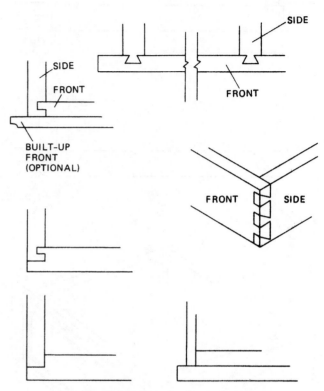

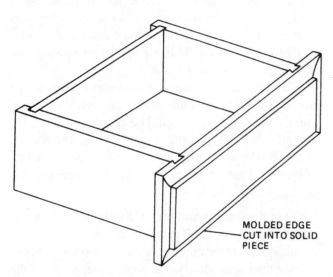

Fig. 7-17 *Lip drawers have a lip that covers the opening in the frame. Drawer fronts can be molded for appearance.*

Fig. 7-19 *Joints for drawer fronts.*

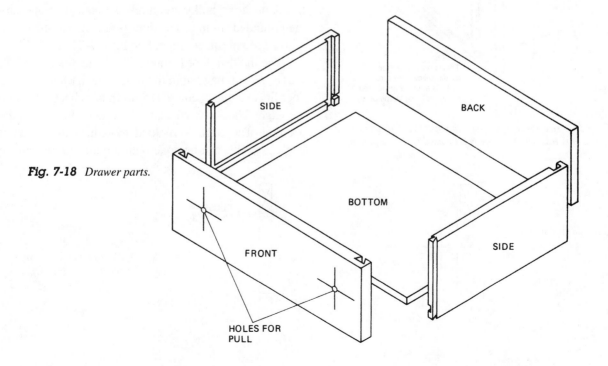

Fig. 7-18 *Drawer parts.*

pensive materials. They are selected for straightness and sturdiness. Plywood is not satisfactory for drawer sides and backs.

The joints for the back are cut into the sides. The joints should be cut on the correct sides. Next, the

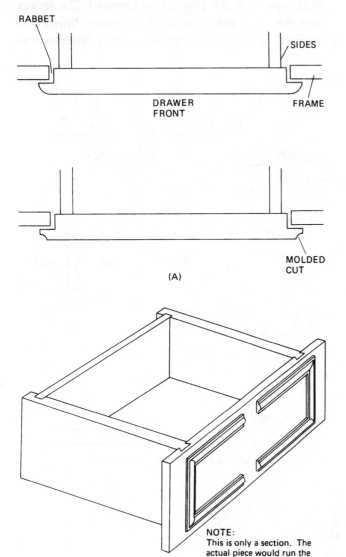

(A)

NOTE:
This is only a section. The actual piece would run the full distance. This is used to show molding shape only.

(B)

Fig. 7-20 (A) These drawer types cover the opening. The accuracy needed for fitting is less. (B) Molding can be used to give a paneled effect.

grooves for the drawers are cut in each side. Be sure that they align properly with the front groove.

Next, the back piece is cut to the correct size. Again, note that the back rests on the bottom. No groove is cut. All pieces are then sanded smooth.

A piece of hardboard or plywood is chosen for the bottom. The front and sides of the drawer are assembled. The final measurements for the bottom are made. Then the bottom piece is cut to the correct size. It is lightly sanded around the edges. The bottom is inserted into the grooves of the bottom and front. This is a trial assembly to check the parts for fit. Next, the drawer is taken apart again. Grooves are cut in the sides for the drawer guides. Also, any final adjustments for assembly are made.

The final assembly is made when everything is ready. Several processes may be used. However, it is best to use glue on the front and back pieces. Sides should not be glued. This allows for expansion and contraction of the materials.

The drawer is checked for squareness. Then one or two small finish nails are driven through the bottom into the back. One nail should be driven into each side as well. This will hold the drawer square. The bottom of the drawer is numbered to show its location. A like number is marked in the cabinet. This matches the drawer with its opening.

Cabinet door construction   Cabinet doors are made in three basic patterns. These are shown in Fig. 7-21. A common style is the rabbeted style. A rabbet about one-half the thickness of the wood is cut into the edge of the door. This allows the door to fit neatly into the opening with a small clearance. It also keeps the cabinet door from appearing bulky and thick. The outside edges are then rounded slightly. The door is attached to the frame with a special offset hinge. See Figs. 7-22 and 7-23.

The molded door is much like the lip door. See Fig. 7-21B. As can be seen in the figure, the back side of the door fits over the opening. However, no rabbet is cut into the edge. This way, no part of the door fits inside the opening. The edge is molded to reduce the apparent thickness. This can be done with a router. Or, it may be

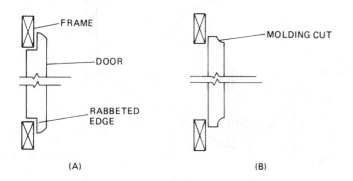

(A)   (B)

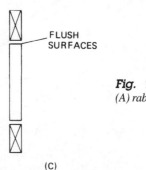

(C)

Fig. 7-21 Cabinet door styles: (A) rabbeted; (B) molded; (C) flush.

Fig. 7-22 *A colorful kitchen with rabbeted lip drawers and cabinets. Note the lack of pulls.* (Armstrong Cork)

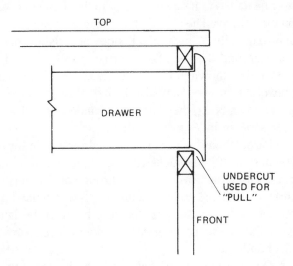

TOP

DRAWER

UNDERCUT
USED FOR
"PULL"

FRONT

Fig. 7-23 *How drawers and doors are undercut so that pulls need not be used.*

done at a factory where the parts are made. This method is becoming more widely used by carpenters. Its advantages are that it does not fit in the opening. That means no special fitting is needed. Also, special hinges are not required. The appearance gives extra depth and molding effects. These are not available for other types of cabinets.

Both styles of door are commonly paneled. Paneled doors give the appearance of depth and contour. Doors may be made from solid wood. However, when cabinets are purchased, the panel door is very common. The panel door has a frame much like that of a screen door. Grooves are cut in the edges of the frame pieces. The panel and door edges are then assembled and glued solidly together. Cross sections of solid and built-up panels are shown in Fig. 7-24.

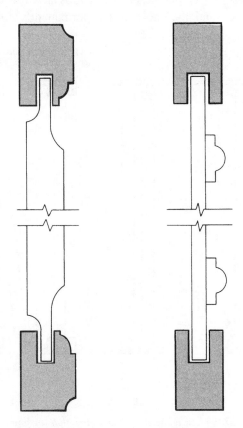

Fig. 7-24 *(A) Cross section of a solid door panel. (B) Cross section of a built-up door panel.*

Flush doors fit inside the cabinet. These appear to be the easiest to make. However, they must be cut very carefully. They must be cut in the same shape as the opening. If the cut is not carefully made, the door will not fit properly. Wide or uneven spaces around the door will detract from its appearance.

Sliding doors Sliding doors are also widely used. These doors are made of wood, hardboard, or glass. They fit into grooves or guides in the cabinet. See Fig. 7-25. The top groove is cut deeper than the bottom groove. The door is installed by pushing it all the way to the top. Then the bottom of the door is moved into the bottom groove. When the door is allowed to rest at the bottom of the lower groove, the lip at the top will still provide a guide. Also, special devices may be purchased for sliding doors.

Making the countertop Today, most cabinet tops are made from laminated plastics. There are many trademarks for these. These materials are usually $\frac{1}{16}$ to $\frac{1}{8}$ inch thick. The material is not hurt by hot objects and does not stain or peel. The laminate material is very hard and durable. However, in a thin sheet it is not strong. Most cabinets today have a base top made from plywood or chipboard. Usually, a ½-inch thickness, or more, is used for countertops made on the site.

Also, specially formed counters made from wood products may be purchased. These tops have the plastic laminates and the mold board permanently formed

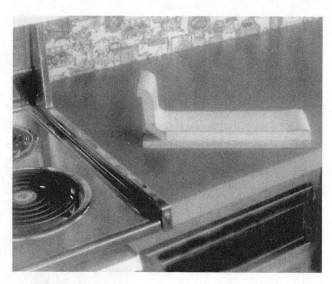

Fig. 7-26 *Countertops may be flat or may include splashboards.*

into a one-piece top. See Fig. 7-26. This material may be purchased in any desired length. It is then cut to shape and installed on the job.

The top pieces are cut to the desired length. They may be nailed to the partitions or to the drawer kickers on the counter. The top should extend over the counter approximately $\frac{3}{8}$ inch. Next, sides or rails are put in place around the top. These pieces can be butted or rabbeted as shown in Fig. 7-27. These are nailed to the plywood top. They may also be nailed to the frame of the counter. Next, they should be sanded smooth. Uneven spots or low spots are filled.

Particleboard is commonly used for a base for the countertop. It is inexpensive, and it does not have any grain. Grain patterns can show through on the finished surface. Also, the grain structure of plywood may form pockets. Glue in pockets does not bond to the laminate. Particleboard provides a smooth, even surface that bonds easily.

Once the counter has been built, the top is checked. Also, any openings should be cut. Openings can be cut for sinks or appliances. Next, the plastic laminate is cut to rough size. Rough size should be $\frac{1}{8}$ to $\frac{1}{4}$ inch larger in each dimension. A saw is used to cut the laminate as in Fig. 7-28.

Next, contact cement is applied to both the laminate and the top. Contact cement should be applied with a brush or a notched spreader. Allow both surfaces to dry completely. If a brush is used, solvent should be kept handy. Some types of contact cement are water-soluble. This means that soap and water can be used to wash the brush and to clean up.

When the glue has dried, the surface is shiny. Dull spots mean that the glue was too thin. Apply more glue over these areas.

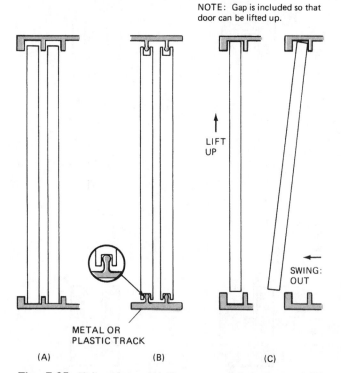

NOTE: Gap is included so that door can be lifted up.

LIFT UP

SWING OUT

METAL OR PLASTIC TRACK

(A)   (B)   (C)

Fig. *7-25* *Sliding doors. (A) Doors may slide in grooves. (B) Doors may slide on a metal or plastic track. (C) To remove, lift up and swing out.*

It usually takes about 15 or 20 minutes for contact cement to dry. As a rule, the pieces should be joined within a few minutes. If they are not, a thin coat of contact cement is put on each of the surfaces again.

To glue the laminate in place, two procedures may be used. First, if the piece is small, a guide edge is put in place. The straight piece is held over the area and the guide edge lowered until it contacts. The entire piece is lowered into place. Pressure is applied from the center of the piece to the outside edges. The hands may be used, but a roller is better.

For larger pieces, a sheet of paper is used. Wax paper may be used, but almost any type of paper is acceptable. The glue is allowed to dry first. Then the paper is placed on the top. The laminate is placed over the paper. Then the laminate is positioned carefully. The paper is gently pulled about 1 inch from beneath the laminate. The position of the laminate is checked. If it is in place, pressure may be applied to the exposed edge. If it is not in place, the laminate is moved until it is in place. Then the exposed edge is pressed until a bond is made. The paper is removed from the entire surface. Pressure is applied from the middle toward the edges. See Fig. 7-29A to M.

Trim for laminated surfaces The edges should be trimmed. The pieces were cut slightly oversize to allow for trimming. The tops should extend over the sides slightly. The tops and corners should be trimmed so that a slight bevel is exposed. This may be done with a special router bit, as in Fig. 7-30A. It may also be done with a sharp and smooth file, as shown in Fig. 7-30B.

The back of most countertops has a raised portion called a splashboard. Splashboards and countertops may be molded as one piece. However, splashboards are also made as two pieces, in which case metal cove and cap strips are applied at the corners. Building codes may set a minimum height for these splashboards. The FHA requires a minimum height of 4 inches for kitchen counters.

Installing hardware Hardware for doors and drawers means the knobs and handles. These are frequently

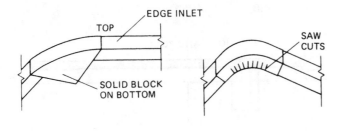

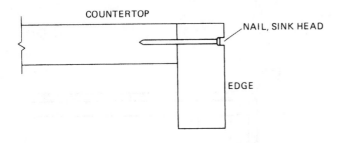

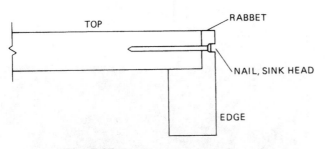

Fig. 7-27 Blocking up counter edges.

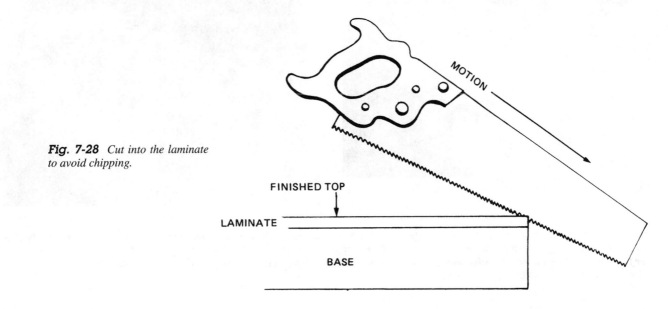

Fig. 7-28 Cut into the laminate to avoid chipping.

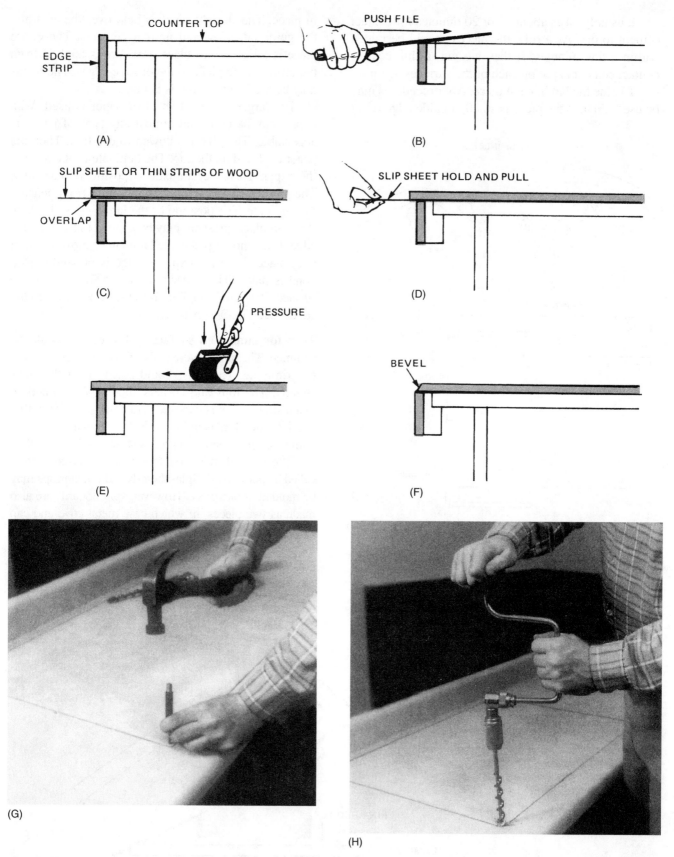

Fig. 7-29 *Applying plastic laminate to countertops. (A) Apply the edge strip. (B) Trim the edge strip flush with the top. (C) Apply glue. Lay a slip sheet or sticks in place when the glue is dry. (D) Position the top. Remove the slip sheet or sticks. (E) Apply pressure from center to edge. (F) Trim the edge at slight bevel. (G) To cut holes for sinks, first center-punch for drilling. (H) Next drill holes at corners.*
(Formica)

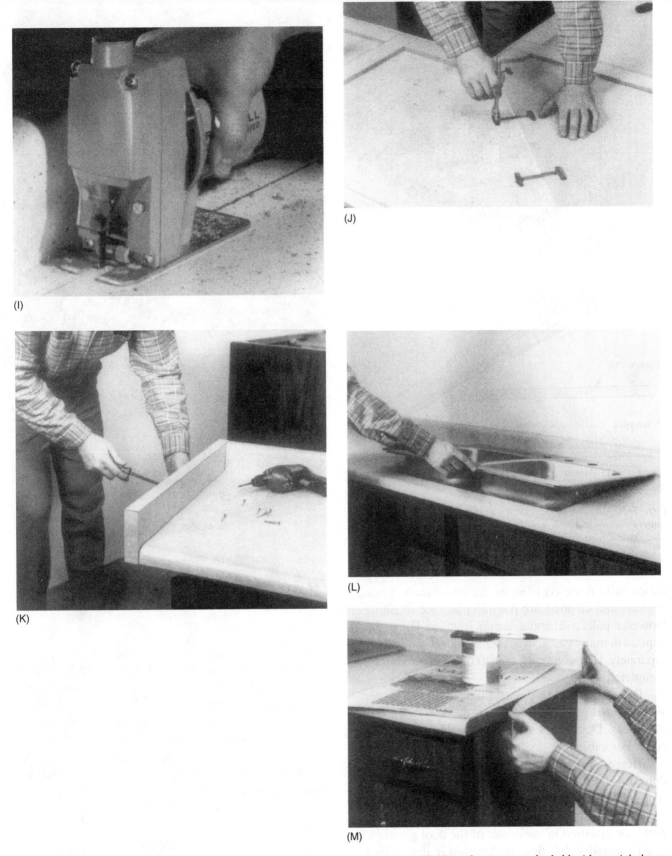

Fig. 7-29 *Continued. (I) Then cut out opening* (Rockwell International, Power Tool Division). *(J) Mitered corners can be held with special clamps placed in cutouts on the bottom* (Formica). *(K) Ends may also be covered by splashboards* (Formica). *(L) Lay sink in opening.* (Formica). *(M) Ends may be covered by strips.* (Formica)

(A)

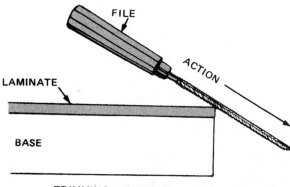

(B)

Fig. 7-30 *(A) Edges may be trimmed with edge trimmers or* ***routers*** *(Rockwell International, Power Tool Division). (B) Edges may also be filed. Note direction of force.*

(A)

(B)

Fig. 7-31 *(A) Pulls may be located in the center for dramatic effect. (B) Regular pull location.*

called *pulls*. A variety of styles are available. As a rule, drawers and cabinets are put into place for finishing. However, pulls and handles are not installed. Hinges are applied in many cases. In others, the doors are finished separately. However, pulls are left off until the finish is completed.

Drawer pulls are placed slightly above center. Wall cabinet pulls are placed in the bottom third of the doors. Door pulls are best put near the opening edge. For cabinet doors in bottom units, the position is different. The pull is located in the top third of the door. It is best to put it near the swinging edge. Some types of hardware, however, may be installed in other places for special effects. See Fig. 7-31A and B. These pulls are installed in the center of the door panel.

To install pulls, first the location is determined. Sometimes a template can be made and used. Whatever method is used, the locations of the holes are found. They are marked with the point of a sharp pencil. Next, the holes are drilled from front to back. It is a good idea to hold a block of wood behind the area. This reduces splintering.

There are other types of hardware. These include door catches, locks, and hinges. The carpenter should always check the manufacturer's instructions on each.

As a rule, it is easier to attach hinges to the cabinet first. Types of hinges are shown in Fig. 7-32. Types of door catches are shown in Fig. 7-33.

Shelves

Most kitchen and bathroom cabinets have shelves. Also, shelves are widely used in room dividers, bookshelves,

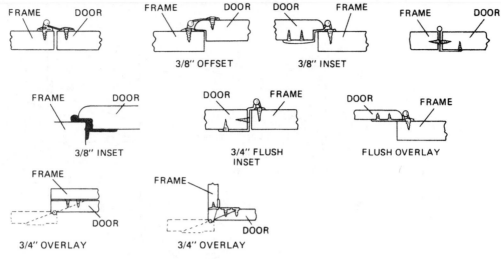

FRAME DOOR

FRAME DOOR
3/8" OFFSET

DOOR FRAME
3/8" INSET

FRAME DOOR

FRAME DOOR
3/8" INSET

DOOR FRAME
3/4" FLUSH INSET

DOOR FRAME
FLUSH OVERLAY

FRAME DOOR
3/4" OVERLAY

FRAME DOOR
3/4" OVERLAY

Fig. 7-32 *Hinge types.*

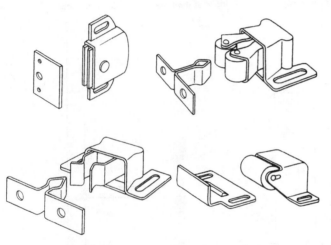

Fig. 7-33 *Types of cabinet door catches.*

and closets. Several methods of shelf construction may be used. Figure 7-34 shows some types of shelf construction. Note that each of these allows the shelf location to be changed.

For work that is not seen, shelves are held up by ledgers. Figure 7-35 shows the ledger method of shelf construction.

Special joints may also be cut in the sides of solid pieces. These are types of dados and rabbet joints. Figure 7-36 shows this type of construction.

As a rule, a ledger-type shelf is used for shelves in closets, lower cabinets, and so forth. However, for exposed shelves, a different type of shelf arrangement is used. Adjustable or jointed shelves look better on bookcases, for instance.

Facing pieces are used to hide dado joints in shelves. See Fig. 7-37. The facing may also be used to make the cabinet flush with the wall. The facing can be inlet into the shelf surface.

Applying Finish Trim

Finish trim pieces are used on the base of walls. They cover floor seams and edges where carpet has been laid. Also, trim is used around ceilings, windows, and other areas. As a rule, a certain procedure is followed for cutting and fitting trim pieces. Outside corners, as in Fig. 7-38, are cut and fit with miter joints. These are cut with a miter box. See Fig. 7-39.

However, trim for inside corners is cut with a different joint. This is done because most corners are not square. Miter joints do not fit well into corners that are not square. Unsightly gaps and cracks will be the result of a poor fit. Instead, a coped joint is used. See Fig. 7-38.

For a coped joint, the first piece is butted against the corner. Then the outline is traced on the second piece. A scrap piece is used for a guide. The outline is cut, using a coping saw. See Fig. 7-40. The coped joint may be effectively used on any size or shape of molding.

APPLYING FINISH MATERIALS

As a rule, the millwork is finished before the wall surfaces. Many wood surfaces are stained rather than painted. This enhances natural wood effects. Woods such as birch are commonly stained to resemble darker woods such as walnut, dark oak, and pecan. These stains and varnishes are easily absorbed by the wall. They are put on first so that they do not ruin the wall finish.

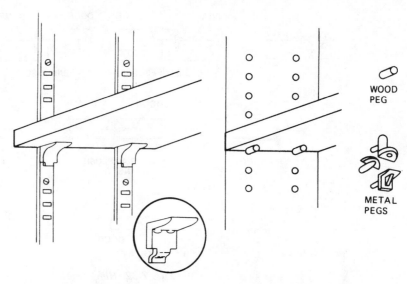

Slotted bookshelf standards and clips
are ideal if you want to adjust bookcase
shelves, but they add to total cost.

If you use wood or metal pegs set into holes, you must be sure to drill
holes at the same level and 3/8 inch deep.

WOOD
PEG

METAL
PEGS

Fig. 7-34 *Methods of making
adjustable shelves.*

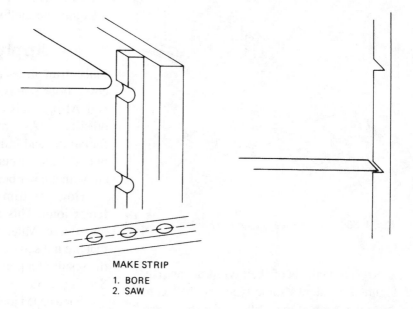

MAKE STRIP
1. BORE
2. SAW

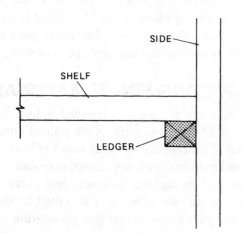

SIDE

SHELF

LEDGER

Fig. 7-35 *Ledgers or cleats are used for shelves and steps where
appearance is not important.*

Paint for wood trim is usually a gloss or semigloss
paint. These paints are more washable, durable, and
costly. Flat or nonglare paints are widely used on
walls.

Paints, stains, and varnishes are often applied by
spraying. This is much faster than rolling. When spray
equipment is used, there is always an overspray. This
overspray would badly mar a wall finish. However, the
wall finish can be put on easily over the overspray. See
Fig. 7-41.

To prepare wood for stain or paint, first sand it
smooth. Mill and other marks should be sanded until
they cannot be seen. Hand or power sanding equipment may be used.

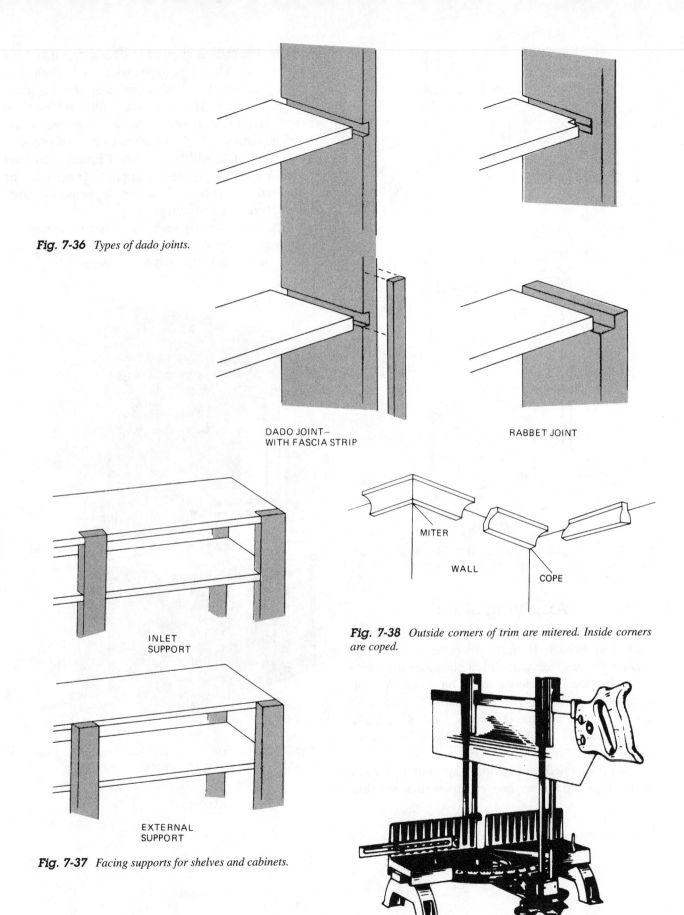

Fig. 7-36 *Types of dado joints.*

DADO JOINT—
WITH FASCIA STRIP

RABBET JOINT

INLET
SUPPORT

EXTERNAL
SUPPORT

Fig. 7-37 *Facing supports for shelves and cabinets.*

MITER

WALL

COPE

Fig. 7-38 *Outside corners of trim are mitered. Inside corners are coped.*

Fig. 7-39 *A miter box used for cutting miters.*

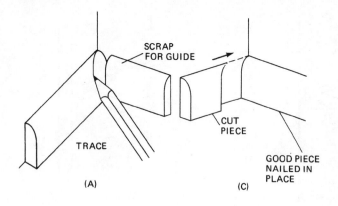

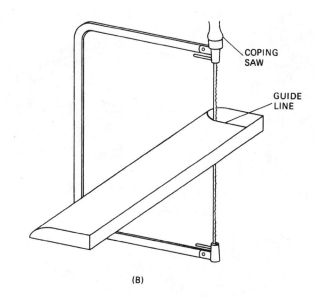

Fig. 7-40 *Making a coped joint. (A) Tracing the outline. (B) Cutting the outline. (C) Nailing the molding in place.*

Applying Stain

Stain is much like a dye. It is clear and lets the wood grain show through. It simply colors the surface of the wood to the desired shade. It is a good idea to make a test piece. The stain can be tested on it first. A scrap piece of the same wood to be finished is used. It is sanded smooth and the stain is applied. This lets a worker check to see if the stain will give the desired appearance.

Stain is applied with a brush, a spray unit, or a soft cloth. It should be applied evenly with long, firm strokes. The stain is allowed to sit and penetrate for a few minutes. Then it is wiped with a soft cloth. It is also a good idea to check the manufacturer's application instructions. There are many different types of stains. Thus there are several ways of applying stains.

The stain should dry for a recommended period of time. Then it should be smoothed lightly with steel wool. Very little pressure is applied. Otherwise, the color will be rubbed off. If this happens, the place should be retouched with stain.

The stain is evenly smoothed. Then varnish or lacquer may be applied with a brush or with a spray gun. Today, most interior finishes are sprayed.

Fig. 7-41 *Cabinets are stained before walls are painted or papered. Note the overspray around the edges.*

8
CHAPTER

Bathrooms

SEVERAL TRENDS INFLUENCE BATHROOM DESIGN and appearance. More bathrooms are being built per home, and they include more open areas, greater use of light, more outdoor views, and more features such as saunas, hot tubs, and spas. Many are expanded to provide extra space within the bath area or even a private terrace, garden patio, or deck. Decks and enclosed decks may include sauna, Japanese furo tub, or spa. See Figs. 8-1 through 8-4.

Unusual materials and combinations of materials are also used on walls, for example, wood, stone, tile, plastic laminates, and wainscotting. Color plays a more important role as well, with traditional antiseptic whites and grays being replaced by bright, chromatic colors sharply contrasting with sparkling whites or the natural textures of wood and stone. Walls, floors, and ceilings now contrast and supplement one another (Fig. 8-5) and often match the decor of nearby rooms.

Fig. 8-1 *New trends in bathrooms include open space and comfort.* (American Olean Tile)

Fig. 8-2 *This bath includes both a spa/tub combination and a glass-enclosed sauna.* (Kohler)

Fig. 8-3 *A spacious bath with luxury and easy maintenance for compartment shower, toilet, bidet, and twin vanities. The adjoining sunken bath overlooks a small garden.* (American Olean Tile)

Fig. 8-4 *The gleaming smooth tile contrasts with the wooden walls.* (American Olean Tile)

Fig. 8-5 *Contrasting tile colors.* (American Olean Tile)

Bathrooms are an essential part of any house. They are designed with various objectives in mind. Some are large; others, small. Some have the minimum of fixtures while others are very elaborate. In this chapter you will learn

- How building codes influence bathroom design
- How plumbing, proper ventilation, and lighting are installed
- How to select the right toilet, tub, and shower
- How to space fittings
- How to select floors, fittings, countertops, and vanities

ROOM ARRANGEMENT

The current trend is toward large bathrooms, but many new homes still use the standard size of 5 feet × 8 feet. These smaller sizes are appropriate for extra bathrooms used mainly by guests or for smaller homes. Size is relatively unimportant if the desired features can be arranged within the space. When you are adding hot tubs, saunas, and so forth, it may be necessary to make projecting bay windows or decks. The key to maximum appearance, utility, and satisfaction is not size, but good arrangement.

In planning room arrangement, the typical pattern of movement in the room should be examined. Avoid major traffic around open doors or other projections. Also consider the number of people who will be using the room at the same time. Many master bathrooms now have two lavatory basins because both members of the couple must rise, dress, and groom themselves at the same time. Having a slightly longer counter with two basins minimizes the frustrations of waiting or trying to use the same basin at the same time.

Entry to the bathroom should be made from a less public area of the house such as a hall. Doors should preferably open into a bathroom rather than out from it. Also, it is a good idea to consider what is seen when the bathroom door is open. The most desirable arrangement is one in which the first view is of the vanity or basin area. Next, a view of the tub or bathing area is appropriate. If possible, the toilet should not be the first item visible from an open door.

Placing new fixtures close to existing lines and pipes minimizes carpentry and plumbing costs. Kitchen plumbing can be located near bathroom plumbing, or plumbing for two bathrooms can be located from the same wall. See Fig. 8-6. Locating the plumbing core in a central area is a good idea.

Water hammer is a pounding noise produced in a water line when the water is turned off quickly. The noise can be reduced by placing short pieces of pipe, called *air traps,* above the most likely causes of quick turnoffs: the clothes washing machine and the dishwasher. Air traps are rather simple to install and can help quiet a noisy plumbing system.

If future changes are anticipated, rough in the pipes needed for the future and cap them off. This way not all the walls will have to be opened later to make complicated connections. Upstairs and downstairs plumbing can be planned to run through the same wall. Again, this makes plumbing accessible and reduces the amount of carpentry and wall work required.

FUNCTION AND SIZE

Where the bathroom is large enough and where two or more people are frequently expected to use it at the same time, compartments provide required privacy and yet accessibility to common areas at all times. See Fig. 8-7. The bath and toilet are frequently set into separate compartments, which allows one to use either area in a degree of privacy while a lavatory or wash basin is used by someone else.

Sometimes, instead of one large bathroom, the space can be converted into two smaller ones, so that more family members can have access at the same time (Fig. 8-8).

There is nothing wrong with small. There are certain advantages to having small bathrooms. They are more economical to build, they can still accommodate several people, and they are much easier to keep clean. Small bathrooms can have all the main features that

Fig. 8-6 *Locating a second bath near the first minimizes the plumbing needs and expense.*

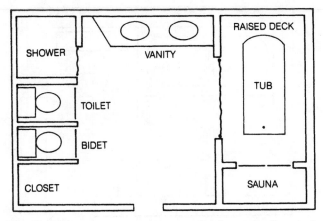

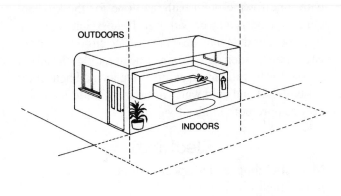

Fig. 8-7 *By putting in compartments for showers, toilets, and bidets, greater privacy results as well as use by more than one person at a time.*

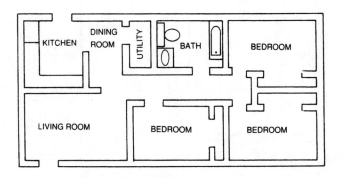

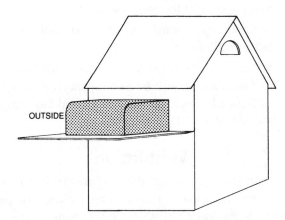

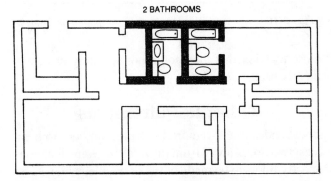

Fig. 8-8 *Two smaller baths can be placed where one larger bath once existed.*

Fig. 8-9 *A cantilever can be used to extend the bathroom for a different result.*

larger bathrooms have by careful use of space through built-in units, vanity counter cabinet space, and pocket doors for compartments. The comfort factors of ventilation, heating and air conditioning, and sound proofing, together with privacy and availability when needed, should be the main consideration.

It is relatively simple to make a small bathroom larger if it is located on an outside wall. Cantilever projections can be used to make windows into large bay windows (Fig. 8-9) or a deck to provide extra space. Also, a bathroom can be enlarged simply by taking other interior space to allow the construction of custom areas,

the addition of privacy gardens or patios, or the addition of spas, saunas, and so forth.

BUILDING CODES

Because bathrooms are complex, building codes may be involved. Codes may designate the types of floors, the materials used, the way things are constructed, and where they are placed. There are usually good reasons for these regulations—even though they might not be obvious. Most cities require rigid inspections based on these codes.

Plumbing

The plumbing is perhaps the most obvious thing affected by codes. Rules apply to the size and type of pipes that can be used, placement of drains, and the placement of shutoff valves.

Shutoff valves allow the water to be turned off to repair or replace fixtures. They are used in two places. The first controls an area. It usually shuts off the cold water supply to different parts of the house. A modern three-bedroom house built over a full basement would typically have three area valves—one for the master bath, one for the main bath, and one for the kitchen. Outdoor faucets may be part of each subsystem based on locations, or they may be on a separate circuit. Second, each fixture, such as a hot water heater, toilet, lavatory, and so forth, will have a cutoff valve located beneath it for both hot and cold water lines. Most building codes now require both kinds of valves.

Electrical

Many building codes specify three basic electrical requirements:

1. The main light switch must be located next to the door but outside the bathroom itself.
2. The main light switch must turn on both the light and a ventilation unit.
3. At least one electrical outlet must be located near the basin, and it must be on a separate circuit from the lights. It should also have a ground fault circuit interrupter (GFCI).

Ventilation

Ventilation is often required for bathrooms. It is a good idea and has many practical implications. In past years, doors and windows were the main sources of bathroom ventilation. They consumed no energy but allowed many fluctuations in room temperature. Forced ventilation is not required if the room has an outside window, but most codes require that all interior bathrooms (those without exterior walls or windows) have ventilation units connected with the lights.

Ventilation helps keep bathrooms dry, to prevent the deterioration of structural members from moisture, rot, or bacterial action. It also reduces odors and the bacterial actions that take place in residual water and moisture.

Fans are vital in humid climates. They should discharge directly to the outdoors, either through a wall or through a roof, and not into an attic or wall space. Ventilation engineers suggest the capacity of the fan be enough to make 12 complete air changes each hour.

Spacing

Building codes may also affect the spacing of fixtures such as the toilet, tub, and wash basin. Figure 8-10 shows the typical spaces required between these units. It is acceptable to have more space, but not less.

The purpose of these codes is to provide some minimum distance that allows comfortable use of the facilities and room to clean them. If there were no codes, some people might be tempted to locate facilities so

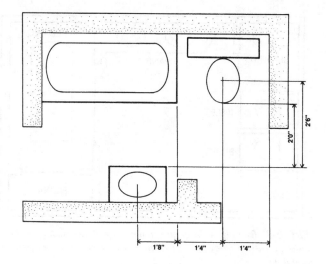

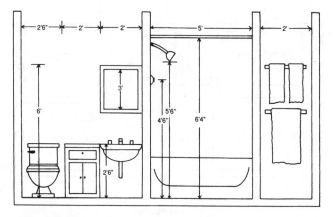

Fig. 8-10 *Typical spacing requirements for bathroom fixtures.*

close together that they could not be safely or conveniently used.

Other Requirements

Local codes might require the bathroom door to be at least two doors away from the kitchen. Some locations specify floors to be made of tile or marble, while others mandate tile or marble thresholds. Certain localities insist that a plastic film or vapor barrier be in place beneath all bathroom floors, and that all basin counters have splash backs or splashboards.

Some specifications might be strict, requiring rigid enclosures on showers or prohibiting the use of glass in shower enclosures. Others might be open, stating the minimum simply to be a rod on which to hang a shower curtain.

FURNISHINGS

Furnishings are the things that make a bathroom either pleasant or drab. They include the fixtures, fittings, and vanity area. The vanity area consists of a lavatory or basin, lights, mirror, and perhaps a counter.

Fixtures

The term *bathroom fixtures* refers to just about everything in the room that requires water or drain connections, such as lavatories (or basins), toilets, bidets, tubs, and showers. Features to consider for each include color, material, quality, cost, and style.

Generally, the better the quality, the higher the cost. Assuming three grades, the cheapest will not be made to withstand long, heavy use. The difference between medium and high quality will be the thickness of the plating and the quality of the exterior finish.

Toilets, bidets, and some lavatories are made of vitreous china, which is a ceramic material that has been molded, fired, and glazed, much as a dinner plate has. This material is hard, waterproof, and easy to clean, and it resists stains. It is very long-lasting; in fact, some china fixtures are still working well after 100 years or more. White is the traditional color, but most manufacturers now provide up to 16 additional colors. Shopping around can give many insights into color and features available.

Toilet Selection

Toilets, or water closets, come in several different mechanisms and styles. The styles include those that fit in corners (Fig. 8-11), rest on the floor, and have their weight supported entirely by a wall. Corner toilets are designed to save space and are particularly functional in very small rooms. Even the triangular tank fits into a corner to save space. The wall-hung toilet is expensive and requires sturdy mounts in the wall. The most common is the floor-mounted toilet (Fig. 8-12).

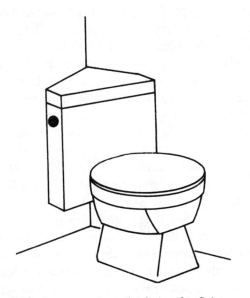

Fig. 8-11 *A space-saving toilet designed to fit into a corner.*

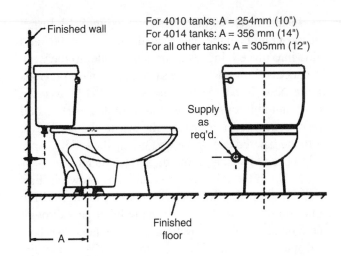

For 4010 tanks: A = 254mm (10")
For 4014 tanks: A = 356 mm (14")
For all other tanks: A = 305mm (12")

Fig. 8-12 *Roughing-in dimensions.* (American Standard)

Floor-mounted toilets can be obtained in different heights. Older people generally find that an 18-inch-high toilet is easier to use than the 14-inch height of conventional units. Heights range from 12 to 20 inches, and some can be purchased with handles and other accessories for the ill or handicapped.

The siphon jet is perhaps the most common type and is most recommended. It is quiet and efficient. Most of the bowl area is covered by water making it easier to clean. The siphon action mechanism is an improvement over the siphon jet. It leaves no dry surfaces, thus making it easier to clean. It is efficient, attractive, and almost silent. It is also the most expensive. Most builders recommend the siphon jet because it costs less.

Up-flush toilets are used in basements when the main sewer line is above the level of the basement floor. These require special plumbing and must be carefully installed.

Toilet Installation

Installing the two-piece toilet requires some special attention to details to prevent leakage and ensure proper operation. The unit itself is fragile and should be handled with care to prevent cracking or breaking. Keep in mind that local codes have to be followed.

Roughing in   Use Fig. 8-12 as a reference. Notice the distance from the wall to closet flange centerline. The distance varies according to the unit selected. For instance, American Standard's 4010 tanks require the distance A to be 254 millimeters (mm) or 10 inches. Model 4014 needs 356 millimeters or 14 inches for distance A. All other tanks require 12 inches or 305 millimeters. The tank should not rest against the wall. Also notice the location of the water supply.

Install the closet bolts as shown in Fig. 8-13. Install the closet bolts in the flange channel, and turn 90° and slide into place 6 inches apart and parallel to the wall. Distance A shown here is the same as that in the previous figure. Next, install the wax seal; see Fig. 8-14. Invert the toilet on the floor (cushion to prevent damage). Install the wax ring evenly around the waste flange (horn), with the tapered end of the ring facing the toilet. Apply a thin bead of sealant around the base flange.

Position the toilet on the flange, as shown in Fig. 8-15. Unplug the floor waste opening and install the toilet on the closet flange so the bolts project through the mounting holes. Loosely install the retainer washers and nuts. The side of washers marked "This side up" *must* face up!

Install the toilet, as per Fig. 8-16. Position the toilet squarely to the wall; and with a rocking motion, press the bowl down fully on the wax ring and flange. Alternately tighten the nuts until the toilet is firmly seated on the floor. CAUTION! Do not overtighten the nuts,

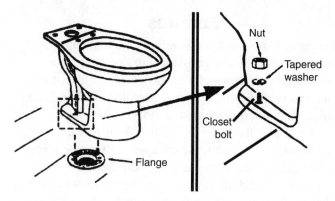

Fig. 8-15 *Positioning the toilet on the flange.* (American Standard)

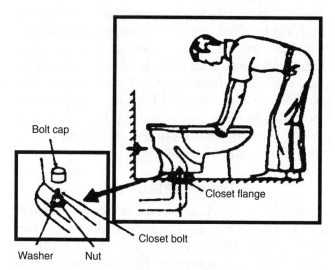

Fig. 8-16 *Installing the toilet.* (American Standard)

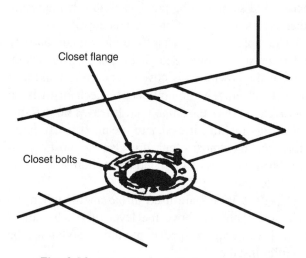

Fig. 8-13 *Closet flange and bolts.* (American Standard)

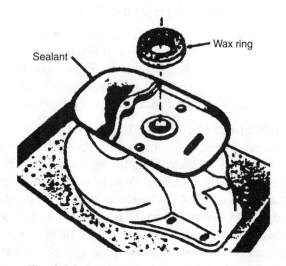

Fig. 8-14 *Installing the wax seal.* (American Standard)

or else the base may be damaged. Install the caps on the washers, and if necessary, cut the bolt height to size before you install the caps. Smooth off the bead of sealant around the base. Remove any excess sealant. Next, install the tank. In some cases, where the tanks and bowl use the Speed Connect System, the tank mounting bolts are preinstalled. Install the large rubber gasket over the threaded outlet on the bottom of the tank, and lower the tank onto the bowl so that the tapered end of the gasket fits evenly into the bowl water inlet opening (See Fig. 8-17) and the tank mounting bolts go through the mounting holes. Secure with metal washers and nuts. With the tank parallel to the wall, alternately tighten the nuts until the tank is pulled down evenly against the bowl surface. CAUTION! Do not overtighten the nuts more than required for a snug fit.

In those instances where the bolts are not preinstalled, start by installing large rubber gaskets over the threaded outlet on the bottom of the tank and then lower the tank onto the bowl so that the tapered end of the gasket fits evenly into the bowl water inlet opening. See Fig. 8-18. Insert the tank mounting bolts and rubber washers from the inside of the tank, through the mounting holes;

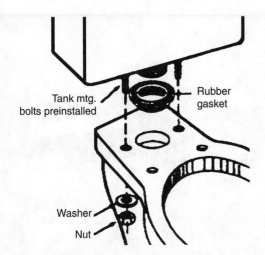

Fig. 8-17 *Installing the tank with preinstalled bolts.* (American Standard)

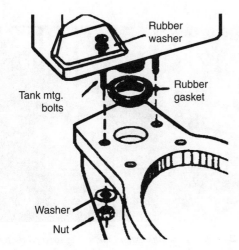

Fig. 8-18 *Installing the tank without preinstalled bolts.* (American Standard)

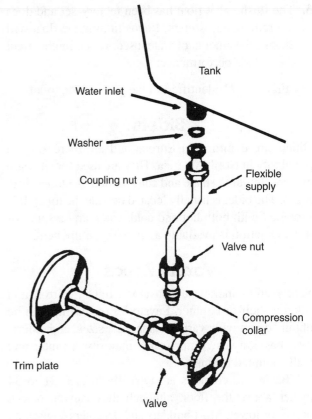

Fig. 8-19 *Connecting to the water supply.* (American Standard)

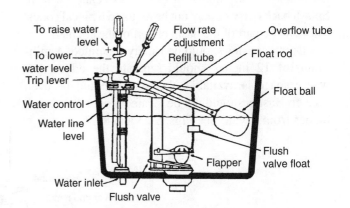

Fig. 8-20 *Making adjustments after installation.* (American Standard)

secure with metal washers and nuts. With the tank parallel to the wall, you can then alternately tighten the nuts until the tank is pulled down evenly against the bowl surface. Again, caution is needed to make sure the nuts are not overtightened. Install the toilet seat according to the manufacturer's directions.

Connect the water supply line between the shutoff valve and tank water inlet fitting. See Fig. 8-19. Tighten the coupling nuts securely. Check that the refill tube is inserted into the overflow tube. Turn on the supply valve, and allow the tank to fill until the float rises to the shutoff position. Check for leakage at the fittings; tighten or correct as needed.

Adjustments There are some adjustments that need to be made in most installations to ensure proper operation. See Fig. 8-20.

1. Flush the tank and check to see that the tank fills and shuts off within 30 to 60 seconds. The tank water level should be set as specified by the mark on the inside of the tank's rear wall.

2. To adjust the water level, turn the water level adjustment screw counterclockwise to raise the level and clockwise to lower the level.

3. To adjust the flow rate (tank fill time), turn the flow rate adjustment screw clockwise to decrease the flow rate. This increases the fill time. Turn the adjustment screw counterclockwise to increase the flow rate or decrease the fill time.

4. Carefully position the tank cover on the tank.

5. The flush valve float has been factory-set and does not require adjustment. Repositioning the float will change the amount of water used, which might affect the toilet's performance.

Figure 8-21 identifies all the parts of the toilet.

Bidets

Bidets are common in Europe and are increasing in popularity in North America. They are used for sitz bath and are similar in shape and construction to a toilet (Fig. 8-22). The bidet is usually located outside the toilet. It is provided with both hot and cold water, and a spray or misting action is available as an extra component.

Vanity Areas

Vanity areas include the lavatory, lights, mirror, and often a shelf or counter. Lavatories, or basins, can be obtained in a wide variety of shapes, sizes, and colors. Two basic styles comprise the majority: counter and wall-mounted basins.

Basin and counters are typically located 31 to 34 inches above the floor, although they can be placed higher or lower. Most builders and designers suggest an 8-inch space between the top of the basin or counter and the bottom of any mirror or cabinet associated with it. Splash backs may or may not be required by local codes. They can be part of the basin or part of the counter.

Double basins should be widely separated. A minimum of 12 inches should separate the edges, but where space is available, this space should be even greater. Basins should not be located closer than 6 to 8 inches from a wall. See Figs. 8-23 and 8-24.

Fig. 8-22 *Toilet and bidet combinations are popular.* (Kohler)

Fig. 8-23 *Double basins extend the vanity area so that it can be used by two people at one time—ideal for working couples.* (American Olean Tile)

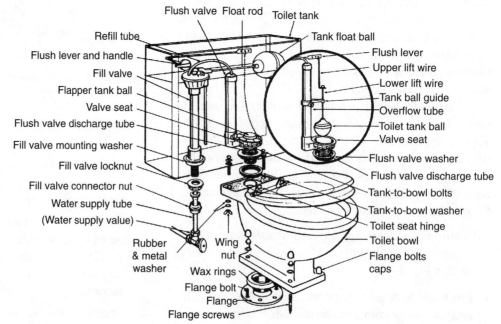

Fig. 8-21 *All parts of the water closet or 1.6-gallon-per-flush, two-piece toilet.*

Refill tube
Flush valve
Float rod
Toilet tank
Tank float ball
Flush lever and handle
Fill valve
Flapper tank ball
Valve seat
Flush valve discharge tube
Fill valve mounting washer
Fill valve locknut
Fill valve connector nut
Water supply tube
(Water supply value)
Rubber & metal washer
Wing nut
Wax rings
Flange bolt
Flange
Flange screws
Flush lever
Upper lift wire
Lower lift wire
Tank ball guide
Overflow tube
Toilet tank ball
Valve seat
Flush valve washer
Flush valve discharge tube
Tank-to-bowl bolts
Tank-to-bowl washer
Toilet seat hinge
Toilet bowl
Flange bolts caps

Fig. 8-24 *A double-basin counter combines crisp lines and easy maintenance.* (Formica)

Countertop Basins

Countertop basin units are popular. See Fig. 8-25. They have one or two basins and frequently incorporate storage areas beneath them. Most basins are designed for counter use and are made from steel coated with a porcelain finish. They can be obtained in traditional white or in a variety of colors that will match the colors of the toilet, tub, and bidet.

Most basins used currently are self-rimming units that seal directly to the countertops for neater, quicker installations. Other styles are available that flush-mount with the countertop and are sealed by a metal or plastic rim, and recessed units that mount below the surface of the counter. Recessed units require greater care to install and are sometimes difficult to keep clean. Note that the self-rimming unit in Fig. 8-26 also has a spray unit for hair care.

Countertop materials Countertops are made from a variety of materials. The most common are probably ceramic tile and plastic laminates. In recent years, the plastic laminates have changed from marblelike patterns to bright, solid colors. While white remains a constant favorite, laminates are available in a wide variety of colors including black, earth tones, and pastels. They can be obtained in a variety of surface finishes and styles, including special countertops with molded splash back and front rims.

Ceramic tile is also an ideal material for counters. It is a hard, durable surface that is waterproof and easy to clean, and it has a beauty that does not fade or wear out. Figure 8-23 features tile counters. Figure 8-24 shows good use of plastic laminates for countertops.

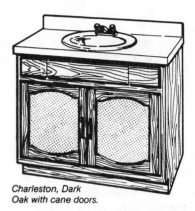

Charleston, Dark Oak with cane doors.

Nova, Honey Oak with planked doors.

Fig. 8-25 *Cabinets topped by one- or two-basin counters are economical and practical.* (NuTone)

Other materials commonly used for countertops include slate, marble, and sometimes wood. Butcher block construction is gaining in popularity and can be either finished with natural oils and waxes or heavily coated with special waterproof plastic finishes. Counters need not be waterproof, but if they are not, water should be wiped up immediately.

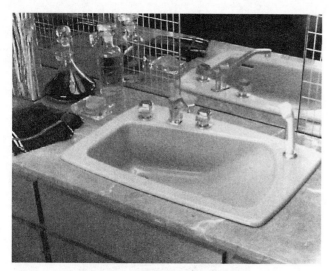

Fig. 8-26 *Self-rimming basins may incorporate hair grooming features.* (Kohler)

Integral tops and basins Counters and basins can be made as one solid piece. The advantage of one-piece construction is that there are no seams to discolor or leak. Both fiberglass and synthetic marble are used for these units. Some provide a complete enclosure for the vanity unit. This protects walls and underlying structures from water damage while being striking to look at.

Wall-Mounted Basins

Wall-mounted basins may be entirely supported by the wall or may be placed to the wall and supported by a pedestal. See Fig. 8-27. Other styles are supported by metal legs at the two front corners. Wall-hung basins placed in corners save space and allow easier movement in smaller bathrooms.

BATHING AREAS

A bathing area may be a tub, a shower, or a combination of both. There are many types and varieties of each, and custom units may be built for all types of baths and combination baths.

Bathtubs

Many people like to soak and luxuriate in a tub. Tubs can be purchased in a variety of shapes and sizes (see Fig. 8-28) and can exactly match the color of the other fittings and fixtures in the bathroom. Tubs can also be purchased in shower-and-tub combinations that match the color of the other fixtures.

Whether to replace an old unit can be an important decision when remodeling. Old tubs can be refinished and built in to provide a newer and more modern appearance. Fittings can be changed and showers added,

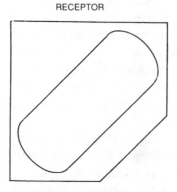

Fig. 8-27 *Pedestal basins are obtainable in a wide variety of shapes and colors.* (Kohler)

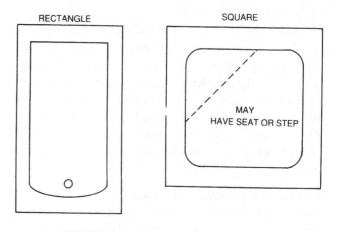

Fig. 8-28 *The three most common tub shapes are rectangle, made to fit into an alcove; square; and receptor. The square and receptor tubs have a longer tub area.*

along with updated wall fixtures such as shelves and soap dishes. Sometimes the antique appearance is preferred, in which case reworking is more desirable than replacing.

Sunken tubs may be standard tubs with special framing to lower them below the surface of the floor. They may be custom-made or may incorporate specially manufactured tubs. Before you install a sunken tub, be sure there is room beneath the bathroom floor. When space below is not available, the alternative is to raise the level of the floor in the remainder of the room. Of course, this presents considerable complications with existing doors and floors. One compromise is to construct a wide pedestal around the lip of the tub (Fig. 8-29). This pedestal can then become a sitting area or a shelf for various articles and can even have built-in storage.

Japanese tubs (*furos*) can be built in and sunken (Fig. 8-30). The tub is simply a deep well that accommodates one or more people on a seat. The seat can be made of wood (the traditional Japanese style), tile, or other material. The soaker is immersed to the neck or shoulders for relaxation. The actual washing with soap or cleaners is traditionally done outside the tub area.

Fig. 8-30 *Furos, or Japanese-style tubs, are becoming increasingly popular.* (American Olean Tile)

Fig. 8-29 *A pedestal can be built around a tub for many reasons. The pedestal can hide pumps, plumbing, and electrical support units.* (Jacuzzi)

Fig. 8-31 *This receptor tub provides a longer than normal bathing area but requires less space and water than a full square tub. It also incorporates a hydromassage unit.* (Kohler)

The standard rectangular tub (Fig. 8-31) is 60 inches long, 32 inches wide, and 16 inches high. It is enclosed or sided on one side, but open at both ends and the remaining side. This shape was designed to fit into an alcove as shown.

Receptor tubs (Fig. 8-31) are squarish, low tubs ranging in height from 12 to 16 inches. Rectangular shapes makes them ideal for corner replacement. They are approximately 36 inches long and 45 inches wide. Square tubs are similar to receptor tubs in that they can be recessed easily into corners and alcoves. Some have special shelves set into corners and others incorporate controls in these areas. Square tubs are approximately 4 to 5 feet square, increasing in 3-inch increments. The receptor tub has a diagonal opening while the square tub may have a truly square basin. The disadvantages of a square tub are that it requires a larger volume of water and that getting in and out is sometimes difficult, particularly for elderly people.

Custom tubs (Fig. 8-32) often feature striking use of ceramic tiles. Tiles can be used in combination with metal, stone, and wood. Specially designed custom showers can have stone and glass walls with low tile sides and are used for sitting and storage.

A spa, also called a hydromassage, whirlpool, water jet, or hot tub, requires extra space to house the jet pump mechanism and the special piping and plumbing. Also, wiring is needed to power the motors. These hydromassage units are available in a variety of sizes and can be incorporated into standard tubs. If they are to be installed in addition to the tub, extra space is required. People with smaller bathrooms find that the combination tub and hydromassage unit (Fig. 8-33) is satisfactory.

Specially shaped tubs can be built by making a frame (Fig. 8-34) that is lined or surfaced with a material such as plywood, which generally conforms to the size and shape desired. Make sure the frame and lining will hold the anticipated weight and movement. Next, tack or staple a layer of fiberglass cloth to the form. Pull

Fig. 8-33 *The hydromassage action can be combined with more conservative settings to look like a conventional tub.* (Jacuzzi)

the cloth to form the shapes around the corners that are desired. If additional support is desired during the shaping process, corner spaces can be filled in and rounded with materials such as fiber insulation. The tub will not need support in the corners, and the fiberglass material itself will be strong enough once completed.

After the cloth has been smoothed to the shape desired (a few seams are all right and will be sanded smooth later on), coat the cloth with a mixture of resin. Tint the resin the color desired for the tub, and use the same color for all coats. Allow the first coat to harden and dry completely. This will stiffen the cloth and give the basic shape for the tub. Next, apply another coat of resin and lay the next layer of cloth onto it. Allow this to harden and repeat the process. At least three layers of fiberglass cloth or fiber will be needed. It is best to add several coats of resin after the last layer of cloth. Three layers of glass fiber are generally applied followed by three more coats of resin. The last three coats of resin are sanded carefully to provide a smooth, curing surface in the exact shape desired.

Showers

Showers may be combined with the tub (Fig. 8-35) or may be separate (Fig. 8-36). In some smaller bathrooms, showers are the only bathing facility. They can be custom-made to fit an existing space, or standard units made from metal or fiberglass may be purchased. Shower stalls are available with the floor, three walls, and sometimes molded ceiling as a single large unit. When made of fiberglass, they are molded as a single integral unit. When made of metal, they are joined by permanent joints or seams.

Fig. 8-32 *Tubs can be custom-built to any size and shape. This tile tub features the same color and style of tile for the tub, walls, counter, and floor.* (American Olean Tile)

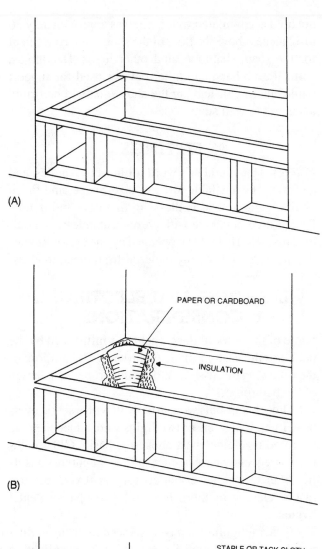

(A)

PAPER OR CARDBOARD

INSULATION

(B)

STAPLE OR TACK CLOTH
TO FRAME AT
TOP AND BOTTOM

LAY CLOTH PATCH
OVER CORNERS

STUFFING IN
CORNERS

(C)

Fig. 8-34 Fiberglass tubs can be custom-built to almost any shape. (A) Wooden frame gives dimension and support. (B) Insulation or cardboard defines the approximate shape and contour. (C) A layer of fiberglass is applied. It can be tinted any color to match decor.

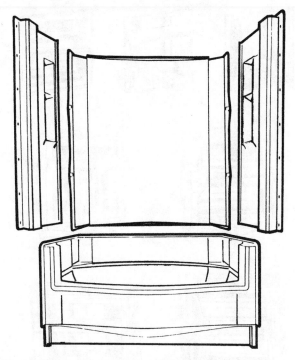

Fig. 8-35 Combination tubs and showers are perhaps the most common unit. They can be made of almost any material or any combination of materials. *(Owens-Corning Fiberglas)*

Fig. 8-36 Showers are often separate compartments for increased privacy.

Standard showers are also available unassembled. This allows a unit to be brought in through halls and doors. The components consist of the drain mechanism, a floor unit, and wall panels.

Manufactured shower units often have handholds, rail ledges for shampoo, built-in soap dishes, and so forth, molded into the walls. Custom-built units can also have conveniences molded into the walls but use separate pieces (Fig. 8-37).

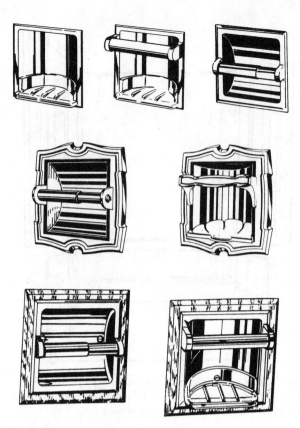

Fig. 8-37 *A wide variety of accessory fittings such as handholds, soap dishes, and storage are available for showers.* (NuTone)

Fiberglass units are usually more expensive than metal ones. Metal units have greater restraints on their design and appearance and are noisier than fiberglass units. Fiberglass should not be cleaned with abrasive cleaners.

Custom-built units are made from a variety of materials including ceramic, tile, wood, and laminated plastics. Tile is an ideal material, but is relatively expensive. The grout between the tiles is subject to stains and is difficult to clean, but special grouts can be used to minimize these disadvantages. Floors for custom-made shower units must be carefully designed to include either a metal drain pan or special waterproofing membranes beneath the flooring.

Using laminated plastics for walls of showers provides several advantages. The material is almost impervious to stains and water, and a variety of special moldings allow the materials to be used. The large size of the panels makes installation quick and easy.

The bottom surface of a shower should have a special nonslip texture; that is, it should be rough enough to prevent skidding, but smooth enough to be comfortable. Neither raised patterns on the bottom of a bathtub nor stick-ons are very good.

Shelves, recessed handholds, and other surfaces in a shower area should be self-draining so that they will not hold accumulated water. Shelves approximately 36 to 42 inches above the floor of the shower are convenient for the soap, shampoo, and other items. Handholds, vertical grab bars, and other devices used for support while entering or leaving the shower should be firmly anchored to wall studs.

FITTINGS

Fittings is the plumber's word for faucets, handles, and so forth. The available array of size, shape, and finish of fittings is almost endless. Both single and double faucets are obtainable with chrome, stainless, or gold-tone finishes. They can be operated by one or two handles that may be made of any material from metal to glass.

LIGHTING AND ELECTRICAL CONSIDERATIONS

Some of the work of the carpenter is influenced by the nature of the custom-made bathrooms. The purpose of the following discussion is to ensure proper installation of the bathroom.

Older bathrooms were usually lit with a single overhead light. Later, one or two lights were added near the mirrored medicine chest above the basin. Older bathrooms frequently have neither sufficient lighting nor sufficient electrical power outlets for hair dryers, electric razors, electric toothbrushes, and water jet for dental hygiene.

General lighting can be enhanced by using ceiling panels or by wall lamps. See Fig. 8-38. Another lighting idea is to use hanging swag lamps. See Fig. 8-39. A large, free-hanging swag lamp would be inappropriate for a small bathroom. General lighting and lighting for the basin areas may be combined for smaller bathrooms.

Basin, or vanity, lamps should be placed above or to the sides of mirrors. They can also be placed in both locations. Light should not shine directly into the eyes but should come from above or the side. One good idea is to use the special "Hollywood" makeup lights around the mirrors (see Fig. 8-40). They eliminate glare and give good light for grooming. They can be a single string of lights above the mirror or can surround the mirror on the sides and top. They are best controlled near the basin area. Separate controls may be desirable so that the user can adjust the lighting and reduce the number of bulbs lit. In addition, dimmer switches vary the intensity.

Special lights may also be desired for radiant heating and for keeping a suntan. The controls for these lights might be housed in several locations, or with the general lighting switch. The controls for special areas

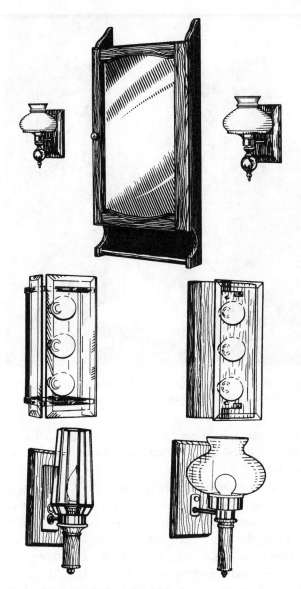

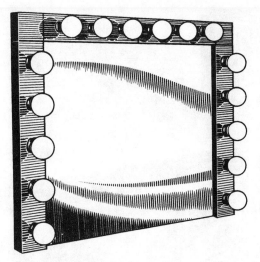

Fig. 8-40 *Strips of lights surrounding a vanity mirror are both useful and popular. This Hollywood style can be controlled by a dimmer switch for increased flexibility.* (NuTone)

such as bathing or toilet compartments are inside the bathroom.

Bathroom outlets should be protected by a GFCI. This is a term used for the *ground fault circuit interrupter.* If you are about to be shocked, it will turn off the circuit.

Newer bathrooms use skylights to give natural light for grooming. They also make good use of picture windows opening onto a patio or deck. Stained glass is also used in some bathroom designs.

The main thing in planning is to avoid a single light source and to use special area lights where needed. Also plan enough outlets to power everything that will be used.

BATHROOM BUILT-INS

A variety of storage space is needed for a bathroom. Tissues and towels are stored for instant use. Also, if several people use a bathroom, more towel space is needed. Clean towels and washcloths, soaps, shampoo, dental articles, grooming items, cleaning equipment, dirty clothes hampers, and even the family linens are all potential storage problems.

Many bathrooms include deep shelves that are often underused because no one can reach the back of them, particularly the top ones. Often the items stacked in the front part of the shelf block the accessibility of the items in back. Sometimes built-ins are simply shelves or drawers that rotate, swing out, or move to allow better use of these back areas that cannot be reached easily. Figure 8-41 shows some ideas to improve efficiency. A closet or cabinet is a must for a bathroom. It should have shelves for a variety of sizes and may incorporate a laundry hamper.

Fig. 8-38 *Specialized lighting for vanity areas can be mounted on walls.* (NuTone)

Fig. 8-39 *Swag lamps are used in larger bathrooms for special lighting effects.* (NuTone)

Fig. 8-41 *Bathroom storage and convenience can be increased by using racks and shelves that swing out or roll out to save shelf space.* (Closet Maid by Clairson)

Fig. 8-42 *A cabinet that extends out from the wall minimizes the strain caused by leaning toward the mirror.* (NuTone)

A laundry chute can be built into the space normally used for a hamper. It will take less space and perhaps save a lot of stair climbing. If a home has a basement laundry, a hole can be cut through the floor deck (but not the frame) so that soiled laundry can be dropped directly into the basement. It is not necessary for the laundry to drop directly into the laundry area, although that would be preferable.

Cabinets beneath the basins are extremely popular because they provide additional storage space housed in an attractive unit. In smaller bathrooms, this becomes more important because there is less space for closets or other built-ins.

People who have basins with wide counters may have to lean forward when using a mirror mounted on the wall. Medicine cabinets with mirrors (Fig. 8-42) hold the storage area out from the wall, which reduces the need to lean.

Where large mirrors are used behind basins, medicine cabinets can be built-in on the ends of the counter. The doors may be either mirrors or wood. Mirrors provide more light and more three-dimensional view while the wood finishes may fit in with the decor. See Fig. 8-43.

Building racks and shelves into the walls is especially helpful when space is at a premium or when storage space is located in an area of high traffic flow. By using the space between the wall studs, items are recessed out of the way. Items such as toothbrushes, water jets, and electric razors can be kept readily accessible, hidden from view, and protected from splashing water by building them into the walls. Electrical outlets can also be built into these areas to power these items. Sliding or hinged doors will hide them from view and keep water and dust off them.

FLOORS AND WALLS

Bathroom floors and walls are subject to water spills and splashes, heat, and high humidity. They should be capable of withstanding the heaviest wear under the most extreme conditions. Good flooring materials for bathrooms include ceramic and quarry tile, stone and brick, wood, resilient flooring, and special carpeting. Good wall materials include stone, tile, wood, and plastic laminates. Walls can also be painted, but regular flat paint is not advised around splash areas such as basins, tubs, and showers. If paint is to be used in those areas, use the best waterproof gloss or semigloss paint.

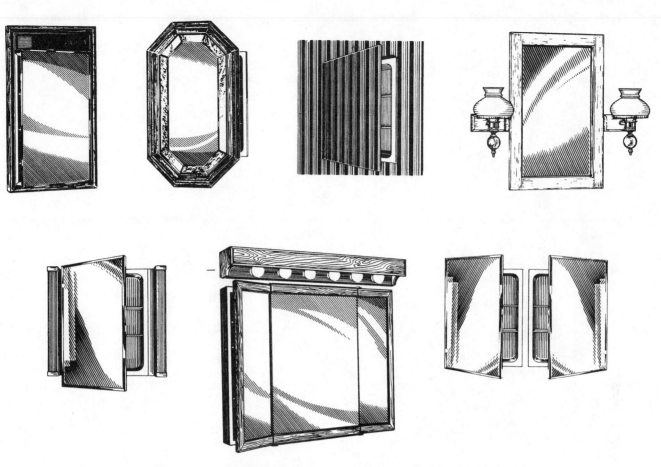

Fig 8-43   *Some examples of the many types of cabinets and doors available.* (NuTone)

9
CHAPTER

Decks, Patios, and Fences

DECKS

Patios can crack and become unsightly in time. They have to be removed and new slabs poured. This can be an expensive job. Most of the damage is done when the fill against the footings or basement walls begins to settle after a few years of snow and rain. One way to renew the usable space that was once the patio is to make a deck. And, of course, if you are building, the best bet is to attach a deck onto the house while it is being constructed. See Fig. 9-1.

There are a few things you should know before replacing the concrete patio slab with a wooden deck. Working drawings for new houses or for remodeling seldom show much in the way of details. The proposed location of any deck should be shown on the floor plan. Its exact size, shape, and construction, however, are usually determined on-site after a study of the site conditions, the contours of the land, orientation of the deck, and the locations of nearby trees. See Fig. 9-2. Decks must be built to the same code requirements as floor framing systems inside the house. Every deck has four parts: a platform, a floor frame, vertical supports, and guardrails.

Platform

The deck itself is built of water-resistant 2 × 4s or 2 × 6s laid flat. All lumber should be of such species as to ensure long duration. That means redwood, cedar, cypress, or pressure-treated wood certified as being suitable for exterior use.

If the decking has vertical graining, the grain runs up and down when the lumber is flat. That means either side may be placed face up. If the decking has flat graining (the grain curves from side to side), each piece must be laid with the annular rings pointing down at the edges. Annular rings are produced as the tree grows. They cause wood to have hard and soft sections. Parts of these rings can be observed at the end of a piece of lumber.

Alternative platform layouts are shown in Figs. 9-3 through 9-7. These decks have the lumber and the necessary connectors for making the deck that particular size. Decking may run parallel to the wall of the house, at right angles to it, or in a decorative pattern, such as a parquet or herringbone. See Fig. 9-8.

Whatever the pattern, each piece of decking must be supported at both ends and long lengths must be given intermediate support. Decking must run at right angles to the framing beneath. To prevent water buildup from rain or snow, the boards are spaced about 1/4 inch apart before they are nailed down. A quick way to do the spacing without having to guess is to put a nail between the decking pieces, so they cannot be pushed together while being nailed.

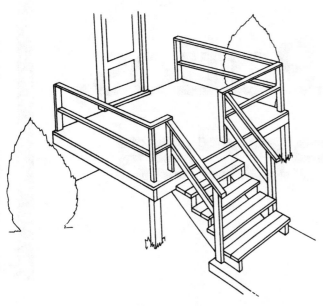

Fig. 9-1 *A deck above grade level.* (Teco)

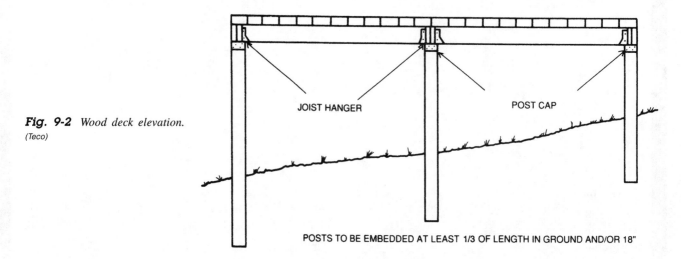

Fig. 9-2 *Wood deck elevation.*
(Teco)

JOIST HANGER   POST CAP

POSTS TO BE EMBEDDED AT LEAST 1/3 OF LENGTH IN GROUND AND/OR 18"

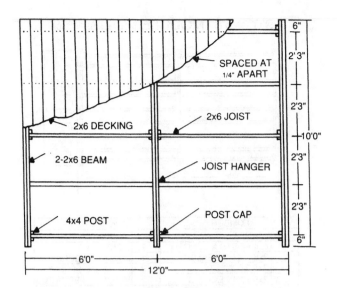

LUMBER		CONNECTORS	
QTY.	TYPE	QTY.	TYPE
9	4x4 POSTS X LENGTH	9	POST CAPS
6	2x6 10/0 BEAMS	20	JOIST HANGERS
10	2x6 6/0 JOISTS		
25☆	2x6 10/0 DECKING	10	FRAMING ANGLES
*	2x6 STAIR TREAD CLEATS	*	3/8" x 4" BOLT
*	2x6 STAIR CARRIAGES	9	5- 1/2" x 6 1/2" PLYWOOD FILLER SHIMS
*	2x6 STAIR TREADS		
*	DETERMINED BY WIDTH & RISE OF STAIR		

☆ IF 2x4 DECKING IS USED QUANTITY REQUIRED IS 39

BASIC WOOD DECK PLAN

Fig. 9-3 *Basic wood deck plan.* (Teco) *Adding to the basic plan can produce a porch. Notice the roof.*

Frame

The deck can be supported by its perimeter on posts or piers. This means the frame will be a simple box-shaped arrangement. When the decking runs parallel to the wall of the house, supporting joists are attached with joist hangers to an edge joist bolted through sheathing into a header. Or, joist hangers can be used to anchor the joist of the deck to the header of the house. See Fig. 9-9. Refer to Figs. 9-4 through 9-7 for alternate ideas of making a deck frame. Single joists can be spaced up to 2 feet 6 inches apart with the joist span no more than 6 feet.

6'0" MODULE (ALL MEMBERS 2x6)

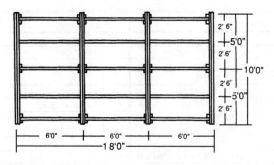

LUMBER		CONNECTORS	
QTY.	TYPE	QTY.	TYPE
12	4x4 POSTS X LENGTH	12	POST CAPS
8	2×6 10/0 BEAMS	30	JOIST HANGERS
15	2×6 6/0 JOISTS	12	5- 1/2" x 6 1/2" PLYWOOD
37☆	2×6 10/0 DECKING		FILLER SHIMS

☆ IF 2x4 DECKING IS USED QUANTITY REQUIRED IS 58

Fig. 9-4 *A deck plan using 6-foot modules.* (Teco)

ADDITION OF 5'0" MODULE
(ALL MEMBERS 2x6)

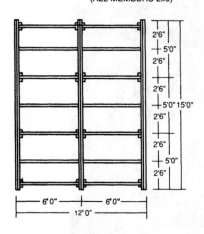

LUMBER		CONNECTORS	
QTY.	TYPE	QTY.	TYPE
12	4x4 POSTS X LENGTH	12	POST CAPS
6	2×6 15/0 BEAMS	28	JOIST HANGERS
14	2×6 6/0 JOISTS	12	5- 1/2" x 6 1/2" PLYWOOD
41☆	2×6 10/0 DECKING		FILLER SHIMS

☆ IF 2x4 DECKING IS USED QUANTITY REQUIRED IS 38

Fig. 9-5 *A deck plan using 5-foot modules.* (Teco)

Support

The other edges of a deck frame are usually supported on 4-foot × 4-foot wood posts. Figure 9-10 shows the post joint assembly detail. Figure 9-11 shows the exterior joist-to-beam connection while the beam-to-post connection is shown in Fig. 9-12. The posts rest on 2-inch wood caps anchored to the tops of concrete piers. Figure 9-13 shows one method used in anchoring. Caps may be omitted if posts are less than 12 inches long. The tops of the piers must be at least 8 inches above grade. If the side is not level, the lengths of posts must be varied so that their tops are at the same level to

9'0" x 5'0" MODULE (ALL MEMBERS 2x6)

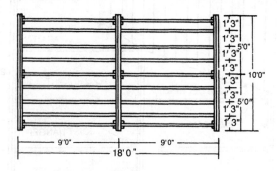

LUMBER		CONNECTORS	
QTY.	TYPE	QTY.	TYPE
9	4x4 POSTS X LENGTH	9	POST CAPS
6	2x6 10/0 BEAMS	36	JOIST HANGERS
18	2x6 9/0 JOISTS	9	5- 1/2" x 6 1/2" PLY-
37☆	2x6 10/0 DECKING		WOOD FILLER SHIMS

☆ IF 2x4 DECKING IS USED QUANTITY REQUIRED IS 58

Fig. 9-6 *A deck plan using 9-foot by 5-foot modules.* (Teco)

8'0" x 4'0" MODULE (ALL MEMBERS 2 x 6)

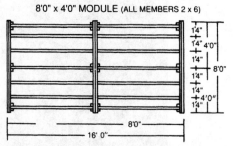

LUMBER		CONNECTORS	
QTY.	TYPE	QTY.	TYPE
9	4x4 POSTS X LENGTH	9	POST CAPS
6	2x6 8/0 BEAMS	28	JOIST HANGERS
14	2x6 8/0 JOISTS	9	5- 1/2" x 6 1/2" PLY -
34☆	2x6 8/0 DECKING		WOOD FILLER SHIMS

☆ IF 2x4 DECKING IS USED QUANTITY REQUIRED IS 52

Fig. 9-7 *A deck plan using 8-foot by 4-foot modules.* (Teco)

support the deck frame. The beams that span from post to post may be 4-inch timbers or a pair of 2-inch joists. The sizes of the beams and the spacing of post have to be considered in the basic design at the beginning. Maximum spans of typical decks are shown in Table 9-1.

As you can see from the table, it is usually less expensive to use larger beams and fewer posts. The posts need more work since they call for a hole to be dug and concrete to be mixed and poured. Plan for the spacing of posts before you begin the job. They have to be placed in concrete and allowed to set for at least 48 hours before you nail to them.

If you space the joists as shown in Fig. 9-3, make sure they correspond to the local code, which should be checked before you start the job. The layouts in Figs. 9-3 through 9-7 are suggestions only and do not indicate compliance with any specific structural code, service, or safety requirements.

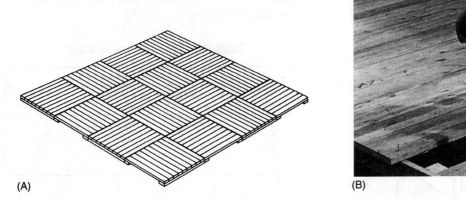

Fig. 9-8 *(A) Parquet deck pattern.* (Western Wood) *(B) Nailing treated southern pine 2 × 4s to a deck attached to the house.* (Southern Forest Products Association)

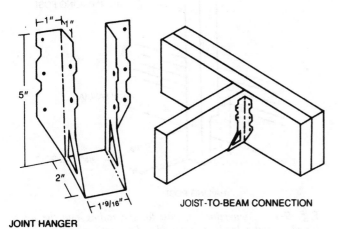

JOIST-TO-BEAM CONNECTION

JOINT HANGER

Fig. 9-9 *Hanger used to connect joists to beam for a deck.* (Teco)

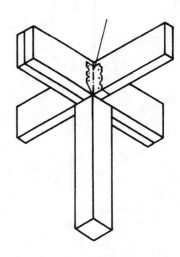

Fig. 9-11 *Beam connection.* (Teco)

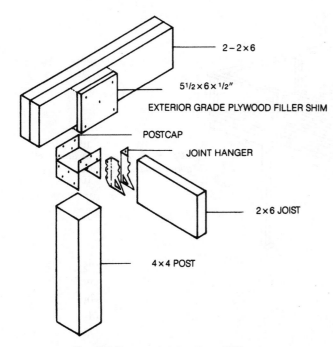

2 – 2 × 6

5¹/₂ × 6 × ¹/₂″

EXTERIOR GRADE PLYWOOD FILLER SHIM

POSTCAP

JOINT HANGER

2 × 6 JOIST

4 × 4 POST

Fig. 9-10 *Post joint detail assembly.* (Teco)

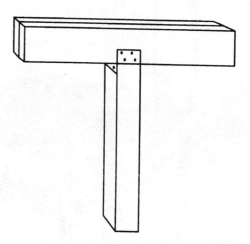

Fig. 9-12 *Beam-to-post connection.* (Teco)

Guardrails

A deck more than 24 inches above the grade must be surrounded with a railing. Supporting posts must be part of the deck structure. The railing cannot be toe-nailed to the deck. The posts may extend upward through the deck, or they may be bolted to joists below deck level. See Fig. 9-14. Spacing of support every 4 feet provides a sturdy railing. Support on 6-foot centers is acceptable. See Fig. 9-15.

Table 9-1 *Deck Beam Spacing*

Deck Width (feet)	Distance Between Posts		
	4 × 6 Beam	4 × 8 Beam	4 × 10 Beam
6	6'9"	9'0"	11'3"
8	6'0"	8'0"	10'0"
10	5'3"	7'0"	8'9"
12	4'6"	6'0"	7'6"

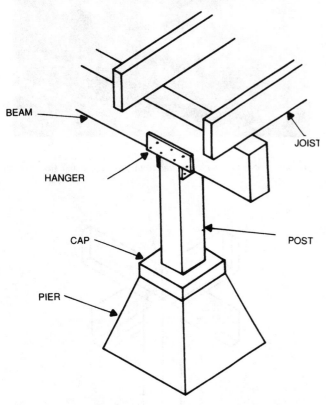

Fig. 9-13 *Wood posts that support the deck rest on caps anchored to concrete piers. Piers may be bought precast, or you can pour your own.*

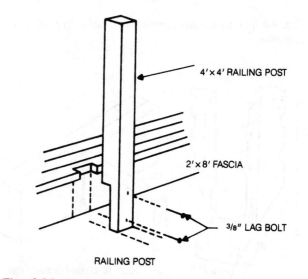

Fig. 9-14 *Connections of post for the railing and the fascia board. Note that lag bolts are used and not nails.* (Western Wood)

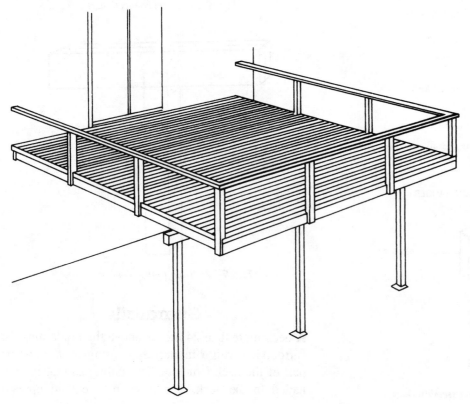

Fig. 9-15 *A deck located well above grade.* (Western Wood)

Making a Hexagonal Deck

A hexagonal deck can add beauty to a house or make it the center of attraction for an area located slightly away from the house. Figure 9-16 shows the basics of the foundation and layout of stringers. Figure 9-17 shows the methods needed to support the deck. To make a hexagonal deck:

1. Lay out the deck dimensions according to the plan and locate the pier positions. Excavate the pier holes to firm soil. Level the bottom of the holes and fill with gravel to raise the piers to the desired height. To check the pier height, lay a 12-foot stringer between piers and check with a level.

2. When all piers are the same height, place a 12-foot stringer in position. Cut and fit 6-foot stringers and toenail them with 10d nails to the 12-foot stringer at the center pier. Use temporary bracing to position the stringers in the correct alignment. Cut and apply the fascia. Cut and apply the stringers labeled B in Fig. 9-16. Note the details in Fig. 9-17.

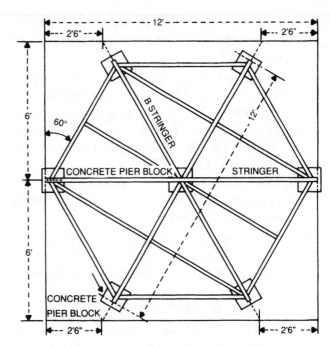

Fig. 9-16 *Layout of hexagonal deck.* (Western Wood)

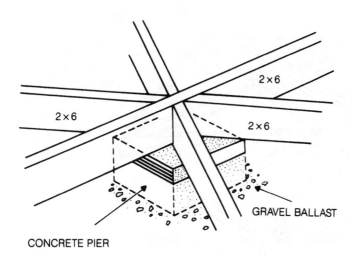

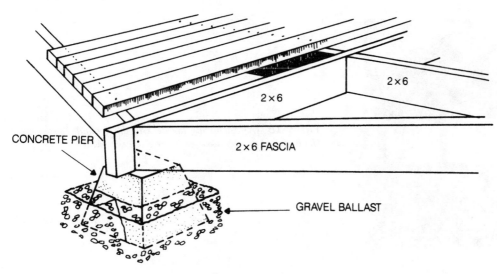

Fig. 9-17 *Supports for the deck.* (Western Wood)

3. Drive the bracing stakes into the ground and nail them to the stringers to anchor the deck firmly into position. See Fig. 9-18.

4. Apply the decking. Begin at the center. See Fig. 9-18. Center edge of the first deck member is over the 12-foot stringer. Use 10*d* nails for spacing guides between deck members. Apply the remaining decking. Nail the decking to each stringer with two 10*d* nails. Countersink the nails. Check the alignment every five or six boards. Adjust the alignment by increasing or decreasing the width between deck members.

5. Tack the trim guide in place and trim the edges, allowing a 2-inch overhang. See Fig. 9-19A. Smooth the edges with a wood rasp or file. See Fig. 9-19B.

Raised Deck

A raised deck (Fig. 9-15) uses a different technique for anchoring the frame to the building. If the deck is attached to a building, it must be inspected by the local building inspector and local codes must be checked for proper dimensions and proper spacing of dimensional lumber. The illustrations here are general and will work in most instances to support a light load and the wear caused by human habitation.

The deck shown in Fig. 9-22 should be laid out according to the sketch in Fig. 9-20. Figure 9-21 shows how the deck is attached to the existing house. Figure 9-22 shows how the decking is installed. Use 10*d* nails for spacing guides between the deck members. Nail the deck member to each stringer with two 10*d* nails. Countersink the nails. Check the alignment every five or six boards. Adjust the alignment by increasing or decreasing the width between the deck members.

The railing posts are predrilled, and then the fascia is marked and drilled for insertion of the $^3/_8$-inch × 3-inch lag bolts per post. The railing cap can be installed with two 10*d* nails per post. This 12- foot × 12-foot deck takes about 5 gallons of Penta or some other type of wood preservative—that is, of course, if you didn't start with pressure-treated wood.

BRACING STAKES FOR ANCHORING TO GROUND

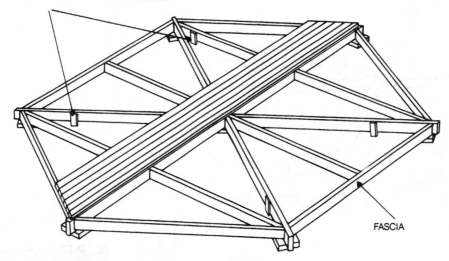

FASCIA

(A)

(B)

Fig. 9-18 *(A) Anchoring the deck to the ground.* (Western Wood) *(B) Finished project.* (Southern Forest Products Association)

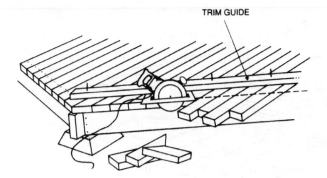

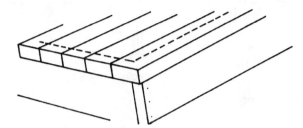

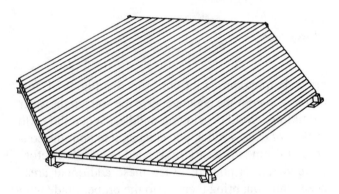

Fig. 9-19A *Trimming the overhand.* (Western Wood)

Fig. 9-19B *The finished hexagonal deck.* (Western Wood)

Steps

If the deck is located at least 12 inches above the grade, you will have to install some type of step arrangement to make it easy to get onto the deck from grade level. Steps will also make it easier to get off the deck.

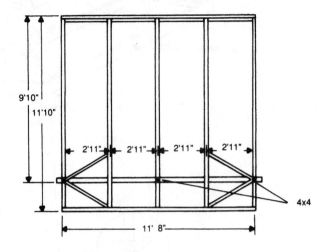

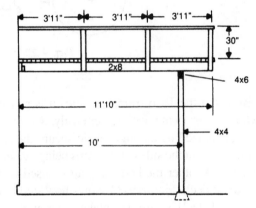

Fig. 9-20 *Supporting the raised 12-foot × 12-foot deck.* (Western Wood)

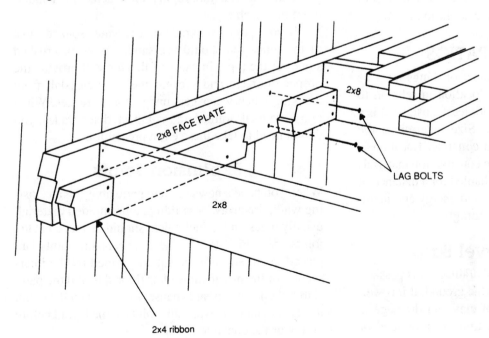

Fig. 9-21 *Notched stringers are used to attach to the nailing ribbon and to the beam.* (Western Wood)

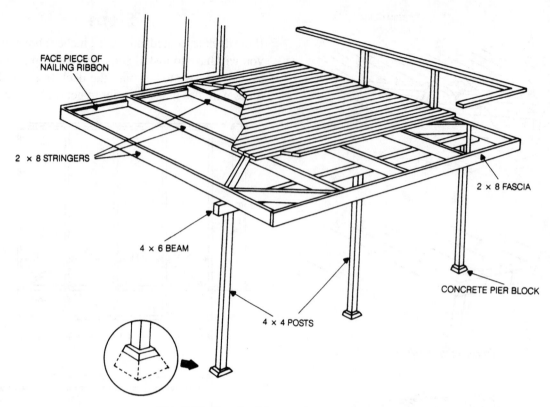

FACE PIECE OF
NAILING RIBBON

2 × 8 STRINGERS

4 × 6 BEAM

4 × 4 POSTS

2 × 8 FASCIA

CONCRETE PIER BLOCK

Fig. 9-22 *Details of the raised deck.* (Western Wood)

By using the Teco framing angles it is possible to attach the steps to the deck rather easily. See Fig. 9-22. These galvanized metal angles help secure the steps to the deck without the sides of the steps being cut and thus weakened. Another method also can be used to attach steps to the deck. Figure 9-23 shows how step brackets (galvanized steel) are used to make a step without having to cut the wood stringers. They come in a number of sizes to fit almost any conceivable application.

CONCRETE PATIOS

Before you pour a patio, a number of things must be taken into consideration: size, location in relation to the house, shape, and last, but not least, cost. Cost is a function of size and location. Size determines the amount of excavation, fill, and concrete. Location is important in terms of getting the construction materials to the site. If materials must be hauled for a distance, it will naturally cost more in time and energy and therefore the price will increase accordingly.

Sand and Gravel Base

If the soil is porous and has good drainage, it is possible to pour the concrete directly on the ground, if it is well tamped. If the soil has a lot of clay and drainage is poor, it is best to put down a thin layer of sand or gravel before pouring the concrete. See Fig. 9-24. Just before you pour the concrete, give the soil a light water sprinkling. Avoid developing puddles. When the earth is clear of excess water, you can begin to pour the concrete. In some instances voids need to be filled before the concrete is poured. When these additional areas need attention, bring them up to the proper grade with granular material thoroughly compacted in a maximum of 4-inch layers.

If you have a subgrade that is water-soaked most of the time, you should use sand, gravel, or crushed stone for the top 6 inches of fill. This will provide the proper drainage and prevent the concrete slab from cracking when water gets under it and freezes. When you have well-drained and compacted subgrades, you do not need to take these extra precautions.

Expansion Joints

When you have a new concrete patio abutting an existing walk, driveway, or building, a premolded material, usually black and ½ inch thick, should be placed at the joints. See Fig. 9-25. These expansion joints are placed on all sides of the square formed by the intersection of the basement wall or floor slab and the patio slab. Whenever a great expanse is covered by the patio, it is best to include expansion joints in the layout before you pour the concrete.

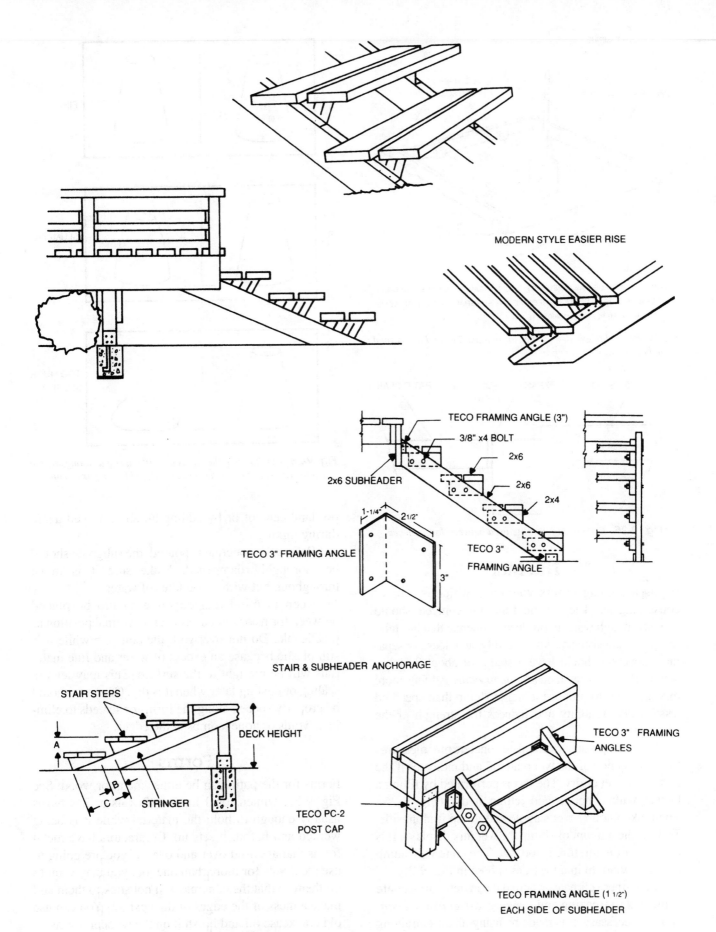

MODERN STYLE EASIER RISE

TECO FRAMING ANGLE (3")

3/8" x4 BOLT

2x6

2x6 SUBHEADER

2x6

2x4

1-1/4"   2 1/2"

TECO 3" FRAMING ANGLE

TECO 3" FRAMING ANGLE

3"

STAIR & SUBHEADER ANCHORAGE

STAIR STEPS

DECK HEIGHT

A

B

C

STRINGER

TECO PC-2 POST CAP

TECO 3" FRAMING ANGLES

TECO FRAMING ANGLE (1 1/2") EACH SIDE OF SUBHEADER

Fig. 9-23 *Methods of mounting steps to a deck* (United Steel Products)

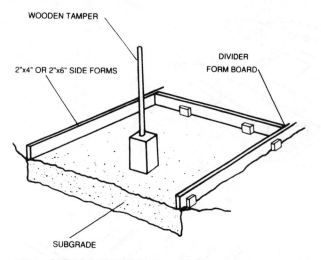

Fig. 9-24 *Using wooden tamper to compact the sand or crushed stone before pouring concrete.*

NOTE: SUBGRADE MAY CONSIST OF CINDER, GRAVEL, OR OTHER SUITABLE MATERIAL WHERE CONDITIONS REQUIRE. THE SUBGRADE SHOULD BE WELL-TAMPED BEFORE PLACING CONCRETE

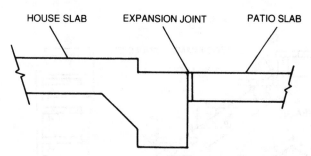

Fig. 9-25 *Expansion joints are used between large pieces.*

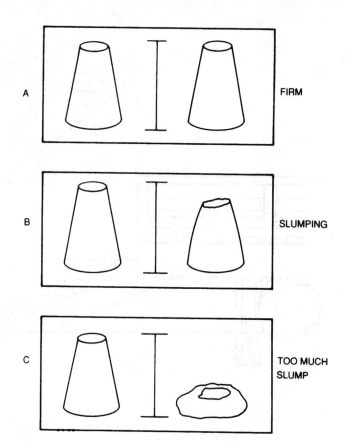

Fig. 9-26 *(A) Slump of the mix is too little. It stands alongside the test cone. (B) The slump is acceptable. (C) The slump is too much.*

The Mix

If you are going to mix your own, which is a time-consuming job, keep in mind that the concrete should contain enough water to produce a concrete that has relatively stiff consistency, works readily, and does not separate. Concrete should have a slump of about 3 inches when tested with a standard *slump cone*. Adding more mixing water to produce a higher slump than specified lessens the durability and reduces the strength of the concrete.

The slump test performed on concrete mix measures the consistency of the wet material and indicates if the mix is too wet or dry. The test is performed by filling a bucket with concrete and letting it dry (Fig. 9-26). Then take another wet sample and dump it alongside. Test for the amount of slump the wet mix displays. It is obvious if the mixture is too wet: The cone will slump down or wind up in a mess, as shown in Fig. 9-26.

In northern climates where flat concrete surfaces are subjected to freezing and thawing, *air-entrained concrete* is necessary. It is made by using an air-entraining portland cement or by adding an air-entrained agent during mixing.

Before the concrete is poured, the subgrade should be thoroughly dampened. Make sure it is moist throughout, but without puddles of water.

Keep in mind that concrete should be placed between forms or screeds as near to its final position as practicable. Do not overwork the concrete while it is still plastic because an excess of water and fine materials will be brought to the surface. This may lead to scaling or dusting later when it is dry. Concrete should be properly spaced along the forms or screeds to eliminate voids or honeycombs at the edges.

Forms

Forms for the patio can be either metal or wood. See Fig. 9-27. Dimensional lumber (2 × 4s or 2 × 6s) is usually enough to hold the material while it is being worked and before it sets up. Contractors have metal forms that are used over and over. If you are going to use the 2 × 4s for more than one job, you may want to oil them so that the concrete will not stick to them and make a mess of the edges of the next job. You can use old crankcase oil and brush it on the wooden forms.

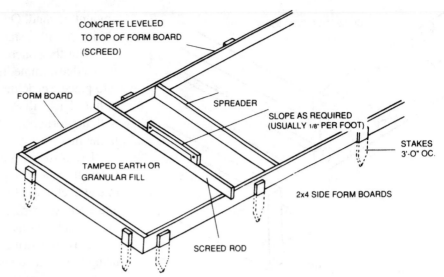

Fig. 9-27 *Forms are made of 2 × 4s or 2 × 6s and staked to prevent movement when the concrete is added and worked.*

CONCRETE LEVELED
TO TOP OF FORM BOARD
(SCREED)

FORM BOARD

SPREADER

SLOPE AS REQUIRED
(USUALLY 1/8" PER FOOT)

STAKES
3'-0" OC.

TAMPED EARTH OR
GRANULAR FILL

2x4 SIDE FORM BOARDS

SCREED ROD

Forms should be placed carefully since their tops are the guides for the screeds. Make sure the distances apart are measured accurately. Use a spirit level to ensure that they are horizontal. If the forms are used on an inclined slab, they must follow the incline. Forms or curved patios or driveways are made from 1/2-inch redwood or plywood. If you want to bend the redwood, soak it in water for about 20 minutes before trying to bend it.

Place stakes at intervals along the outside of the forms and drive them into the ground. Then nail the stakes to the forms to hold them securely in place. The tops of the stakes must be slightly below the edge of the forms so they will not interfere with the use of the strike-off board for purposes of screeding later.

Placing the Joints

Joints are placed in concrete to allow for expansion, contraction, and shrinkage. It is best not to allow the slab to bond to the walls of the house, but to allow it to move freely with the earth. To prevent bonding, use a strip of rigid waterproof insulation, building paper, polyethylene, or something similar. Also use the expansion joint material where the patio butts against a walk or other flat surface. See Fig. 9-28.

Wide areas such as a patio slab should be paved in 10- to 15-foot-wide alternate strips. A construction joint is made by placing a beveled piece of wood on the side forms. See Fig. 9-29. This creates a groove in the slab edges. As the intermediate strips are paved, concrete fills the groove and the two slabs are keyed together. This type of joint keeps the slab surfaces even and transfers the load from one slab to the other when heavy loads are placed on the slabs.

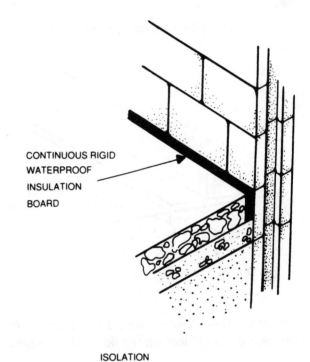

CONTINUOUS RIGID
WATERPROOF
INSULATION
BOARD

ISOLATION

Fig. 9-28 *Isolation joint prevents slab from cracking wall as it expands.*

You have seen contraction joints, often called *dummy joints,* cut across a slab (Fig. 9-30). They are cut to a depth of one-fifth to one-fourth the thickness of the slab. This makes the slab weaker at this point. If the concrete cracks due to shrinkage or thermal contraction, the crack usually occurs at this weakened section. In most instances, the dummy joint is placed in the concrete after it is finished off. A tool is drawn through the concrete before it sets up. It cuts a groove in the surface and drops down into the concrete about 1/2 inch. In case the concrete cracks later after drying it

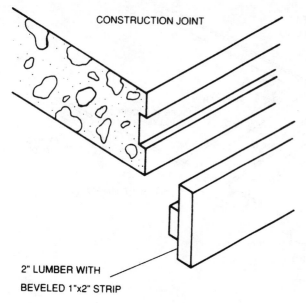

Fig. 9-29 *Construction joint.*

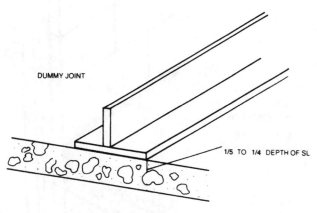

Fig. 9-30 *Dummy or contraction joint.*

will usually follow these grooves. These grooves are usually placed 10 to 15 feet apart on floor slabs or patios.

Pouring the Concrete

Most concrete is ordered from a ready-mix company and delivered with the proper consistency. If you mix your own, some precautions should be taken to ensure its proper placement. Concrete should be poured within 45 minutes of the time it is mixed. Some curing begins to take place after that, and the concrete may become too thick to handle easily. If you are pouring a large area, mix only as much as you can handle within 45 minutes. The first batch should be a small one that you can use for trial purposes. After the first batch, you can determine whether the succeeding batches require more or less sand. Remember, do not vary the water for a thicker or thinner concrete—vary only the

amount of sand. Once you have established the workability you like, stay with that formula.

Pour the concrete into forms so it is level with the form edges. Immediately after the first batch is poured into place, spade the concrete with an old garden rake or hoe. This will even out the wet concrete so it is level with the form boards. Make sure there are no air voids in the mix.

Use a striker board to level off the concrete in the form. It usually requires two people to do this job— one at each end of the board. The idea is to get a level and even top surface on the poured concrete by using the board forms as the guide.

Draw the striker board across the concrete while using the form edges as the guides. Then you can see-saw the board as you move it across. Now it is obvious why the stakes for the form had to be below the edges of the form boards: They should not interfere with the movement of the strike board. The strike board takes off the high spots and levels the concrete. If the board skips over places where it is low and lacks concrete, fill in the low spots and go over them again to level off the new material.

Once you have finished striking the concrete and leveling it off, you may notice a coat of water or a shiny surface. This may not be evenly distributed across the top, but wait until this sheen disappears before you do any other work on the concrete. This may take an hour or two; the temperature determines how quickly the concrete starts to set up. The humidity of the air and the wind are also factors. The concrete will begin to harden and cure. See Fig. 9-31.

Finishing

When the water sheen and bleed water have left the surface of the concrete, you can start to finish the surface of the slab. This may be done in one or more ways, depending on the type of surface you want. Keep in mind that you can overdo the finishing process.

You can bring water under the surface up to the top to cause a thin layer of cement to be deposited near or on the surface. Later, after curing, it becomes a scale that will powder off with use.

Finishing can be done by hand or by rotating power-driven trowels or floats. The size of the job will determine which to use. In most cases, you will have to rent the power-driven trowel or float. Therefore, economy of operation becomes a factor in finishing. Can you do the job by yourself, or will you need help to get it done before the concrete sets up too hard to work? You should make up your mind before you start so you

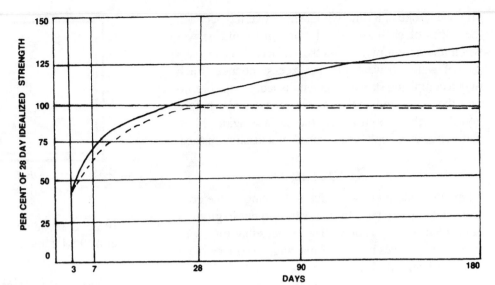

Fig. 9-31 *Relative concrete strength versus curing method.*

can have additional help handy or have the power-driven machines around to speed up the job.

The type of tool used determines the type of finish on the surface of the patio slab. A wood float puts a slightly rough surface on the concrete. A metal trowel or float produces a smooth finish. Extra-rough surfaces are produced by using a stiff-bristled broom across the top.

Floating

A float made of a piece of wood with a handle for use by hand is used to cause the concrete surface to be worked. In some cases it has a long handle so the concrete can be worked by a person standing up and away from the forms. See Fig. 9-32. A piece of plywood or other board can be used to kneel on while floating the surface. The concrete should be set up sufficiently to support the person doing the work. Floating has some advantages. It embeds the large aggregate (gravel) beneath the surface and removes slight imperfections such as bumps and voids. It also consolidates the cement near the surface in preparation for smoother finishes. Floating can be done before or after edging and grooving. If the line left by the edger and groover is to be removed, floating should follow the edging and grooving operation. If the lines are to be left for decorative purposes or to provide a crack line for later movement of the slab, edging and grooving will have to follow the floating operation.

Troweling

Troweling produces a smooth, hard surface. It is done right after floating. For the first troweling, whether by hand or power, the trowel blade must be kept as flat against the surface as possible. If the trowel blade is tilted or pitched at too great an angle, an objectionable washboard or chatter surface will be produced. For first troweling, do not use a new trowel. An older one that has been broken in can be worked quite flat without the edges digging into the concrete. The smoothness of the surface can be improved by a number of trowelings. There should be a lapse of time between successive trowelings to allow the concrete to increase its set or become harder. As the surface stiffens, each successive troweling should be made by a smaller trowel. This gives you sufficient pressure for proper finishing.

Brooming

If you want a rough-textured surface, you can score the surface after it is troweled. This can be done by using a broom. Broom lines should be straight lines, or they can be swirled, curved, or scalloped for decorative purposes. If you want deep scoring of the surface, use a wire-bristle broom. For a finer texture you may want to use a finer-bristle broom. Draw the broom toward

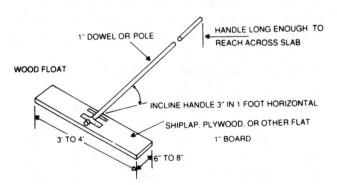

Fig. 9-32 *Construction details for a long-handled float.*

you one stroke at a time with a slight overlap between the edges of each stroke. The broom should be wet when it is first drawn across the surface. You can use a pail of water to wash off the excess concrete. Each time a completed stroke is accomplished, dip the broom into the water and shake off the excess water before you draw the broom across the surface again. Most patios, however, look better with a smooth, silky surface.

Grooving

To avoid random cracking due to heaving of the concrete slab after it has cured, it is best to put in grooves after the troweling is done. Then cracks that form will follow the grooves instead of marring the surface of an otherwise smooth finish.

FENCES

Installing a fence calls for the surveyor's markers to be located in the corners of the property. Lot markers are made of ³/₄-inch pipe, or they could be a corner concrete monument. Many times they are buried as much as 2 feet under the surface. The plot plan may be of assistance in locating the surveyor's markers. Locating proper limits will make sure the fence is on the land it is intended for and not on the property next door.

The local zoning board usually has limitations on the height of the fence and just where it can be placed on a property. If ordinances allow it to be placed exactly on the property line, be sure to check with the abutting neighbors and obtain their consent in order to prevent any future lawsuits.

In most states it is generally understood that if the posts are on the inside facing your property and the fence is on the other side, except for post-and-rail fencing, the fence belongs to you. Local officials will be more than happy to help you in regard to the details.

Installation

Most fences are variations of a simple post, rail, and board design. The post and rail support structure is often made of standard dimension lumber while the fence boards come in different shapes and sizes giving the fence its individual style.

On a corner of the lot, place a stake parallel to the surveyor's marker and attach a chalk line to the stake. Pull the chalk line taut and secure it at the opposite stake. Repeat until all boundaries are covered. See Fig. 9-33. Repeat the procedure to measure the required setback from the original boundary and restake accordingly. You are now ready to install the fence.

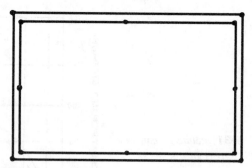

Fig. 9-33 *Layout of the lot and location of posts for the fence.*

To get the fencing started correctly, start from one corner of the lot and place the first post in that corner.

Setting Posts

Setting posts requires patience and is the most critical aspect of fence building. Posts must be sturdy, straight, and evenly spaced for the fence to look and perform properly. Redwood, western cedar, or chemically treated pine can be used for the posts. Posts are commonly placed 8 feet on center. Mark the post locations with stakes.

Dig holes about 10 inches in diameter with a post-hole digger. Holes dug with a shovel will be too wide at the mouth to provide proper support. Auger-type diggers work well in rock-free earth. If you are likely to encounter stones, use a clamshell type of digger.

Set corner posts first. String a line between corner posts to mark the fenceline and align the inside posts. For a 5- or 6-foot fence, postholes should be at least 2 feet deep. A 3-foot hole is required for an 8-foot fence. See Fig. 9-34.

Proper drainage in postholes eliminates moisture and extends the life of posts. Fill the bottom of holes with gravel. Large, flat base stones also aid drainage. Fill in with more gravel, 3 or 4 inches up the post.

For the strongest fence, set the posts in concrete. You may figure your concrete requirements at roughly an 80-pound bag of premixed concrete per post. If you extend the concrete by filling with rocks or masonry rubble, be sure to tap it down. Cleats or metal hardware can be used to strengthen fence posts. Cleats are 2 × 4 wood scraps attached horizontally near the base of the post that provide lateral stability. Large lag screws or spikes partially driven into the post can be used to do the same job when posts are set in concrete.

Make sure the concrete completely surrounds the post to make a collar. It doesn't have to be a full shell. This reduces the tendency to hold water and promote decay. See Fig. 9-35. When you have placed the posts

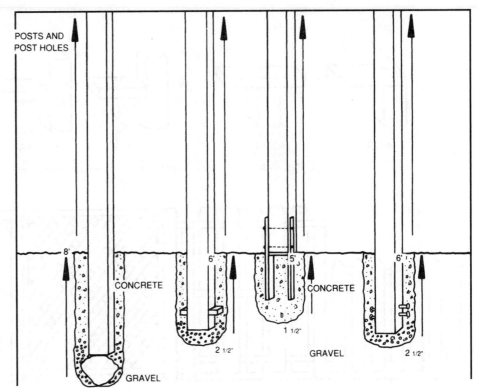

Fig. 9-34 Depths needed for mounting fence posts. (California Redwood Association)

POSTS AND POST HOLES

CONCRETE

GRAVEL

CONCRETE

GRAVEL

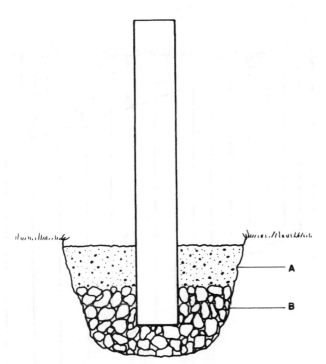

Fig. 9-35 Location of concrete collar around the post at A. The gravel fill is shown at B.

to the height desired, pour the concrete collar around the post. Be sure the posts are perpendicular to the surface and allow the concrete to cure at least 48 hours before you attach the fencing.

Attaching the Rails

Two or three horizontal rails run between the posts. The number of rails depends on the fence height. Rails 8 feet long are common because this length of 2 × 4 is readily available and provides enough support for most styles of fenceboards. Upper rails should rest on the posts. Rails can be butt-joined and mitered at corners. Bottom rails can be toenailed into place, but the preferred method is to place a block underneath the joint for extra support. Metal hardware such as L brackets can be used to secure rails, but make sure all metal fasteners including nails are noncorrosive. See Fig. 9-36.

Attaching Fence Boards

Nailing the fence board in place is the easiest and most satisfying part of the project. Boards 1 inch in thickness and 4, 6, 8, or 10 inches in width are common. The dimensions of the fence boards and the way they are applied will give the fence its character.

There are several creative ways to do this. For fence boards of widths 4 inches and less, use one nail per bearing. For wider fence boards use two nails per bearing. Do not overnail.

Picket fences leave plenty of room between narrow fence boards. They are often used to mark boundaries and provide a minimal barrier. Just enough to keep small children or small dogs in the yard, they are typically 3 or 4 feet high. See Fig. 9-37.

TOP RAILS   BOTTOM RAILS

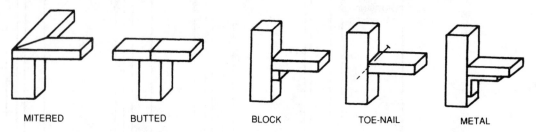

MITERED   BUTTED   BLOCK   TOE-NAIL   METAL

Fig. 9-36 *Attaching rails to the posts.* (California Redwood Association)

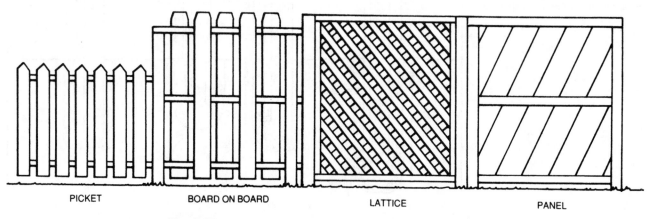

PICKET   BOARD ON BOARD   LATTICE   PANEL

Fig. 9-37 *Various types of fence designs.* (California Redwood Association)

Board on board fences are taller and provide more of a barrier. Distinguished by the fence boards alternating pattern, board on board fences look the same from either side. This allows great flexibility in design and function. Depending on the placement of fence boards, this fence will block the wind and the view.

Lattice panels made from redwood lath or $\frac{1}{2} \times 2$s can be used to create a lighter, more delicate fence. Lattice panels can be prefabricated within a 2×4 frame, then nailed to the posts.

Panel fences create a solid barrier. Panels are formed by nailing boards over rails and posts. By alternating the side that the fence boards are nailed to at each post, you can build a fence that looks the same from both sides.

Stockade fencing is built level. See Fig. 9-38. However, not all ground is flat. You can adjust for the slope of the ground working downhill if possible. Fasten the section into a post that is plumb, then exert downward pressure on the end of the section before you attach it to the post. See Fig. 9-39. This is called *racking*. When finished, all posts should follow the slope of the ground.

Nails and Fasteners

Use noncorrosive nails with redwood outdoors. Stainless steel, aluminum, and top-quality, hot dipped galvanized

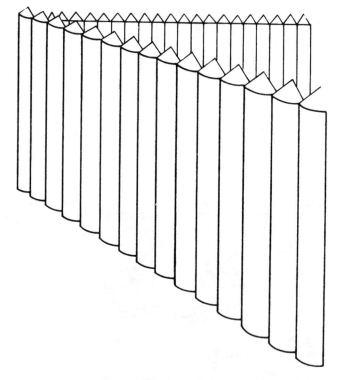

Fig. 9-38 *Stockade fence.*

nails will perform without staining. Inferior hardware, including cheap or electroplated galvanized nails, will corrode and cause stains when in contact with moisture.

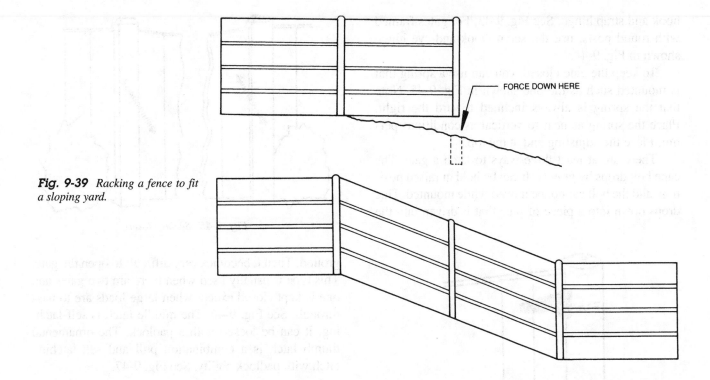

Fig. 9-39 *Racking a fence to fit a sloping yard.*

If you use redwood, do not finish off with varnishes and clear film finishes, oil treatments, or shake and shingle-type paints.

Gates

Getting the gate to operate properly is another problem with fencing. By using the proper hardware and making sure the gate posts are plumb and sturdy, it is possible to install a gate that will work for years. See Fig. 9-40. To apply the hinges, position the straps approximately 4 inches from both the top and the bottom of the gate on the side to be hinged. See Fig. 9-41. Position the gate in the opening and allow adequate clearance between the bottom of the gate and the ground. You can use a wooden block under the gate to ensure proper clearance and aid in the installation. Mark and secure the vertical leaves to the fence posts. Drill pilot

holes for the screws. See Fig. 9-42. For flat gates to be mounted on the fence post, use an ornamental screw

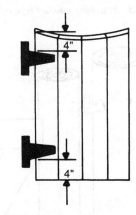

Fig. 9-41 *Allow 4 inches for the hinges.*

Fig. 9-40 *Fence gate spring closer.*

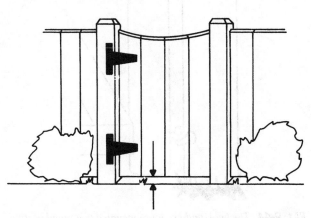

Fig. 9-42 *Place a spacer block of wood under the gate while you attach the hinges.*

hook and strap hinge. See Fig. 9-43. For gates framed with round posts, use the screw hook-and-eye hinge shown in Fig. 9-44.

To keep the gate closed, you can use a spring that is mounted such as the one shown in Fig. 9-45. Note that the spring is always inclined toward the right. Place the spring as near to vertical as conditions permit. Place the adjusting end at the top.

There are at least three ways to latch a gate. The cane bolt drops by gravity. It can be held in raised position, and the bolt cannot be moved while mounted. This drops down into a piece of pipe that is driven into the

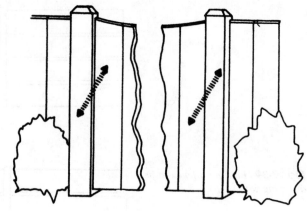

Fig. 9-45 *Spring closers.*

ground. Then it becomes very difficult to open the gate. This type is usually used when there are two gates and one is kept closed except when large loads are to pass through. See Fig. 9-46. The middle latch is self-latching. It can be locked with a padlock. The ornamental thumb latch is a combination pull and self-latching latch with padlock ability. See Fig. 9-47.

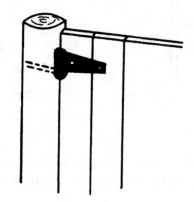

Fig. 9-43 *Attaching another type of hinge.*

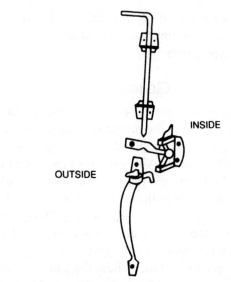

INSIDE

OUTSIDE

Fig. 9-46 *Gate hardware.*

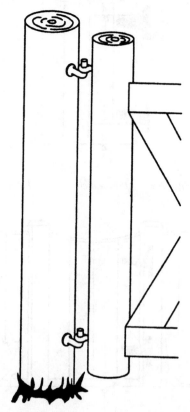

Fig. 9-44 *The hook-and-eye hinge mounted to a round post.*

Fig. 9-47 *Completed installation.*

Index

ABOUT THE AUTHORS

Mark R. Miller is Chairman and Associate Professor of Industrial Technology at Texas A&M University in Kingsville, Texas. He teaches construction courses for future middle managers in the trade. He is coauthor of several technical books, including the best-selling *Carpentry & Construction,* now in its fourth edition. He lives in Kingsville, Texas.

Rex Miller is Professor Emeritus of Industrial Technology at State University College at Buffalo and has taught technical curriculum at the college level for more than 40 years. He is the coauthor of the best-selling *Carpentry & Construction,* now in its fourth edition, and the author of more than 75 texts for vocational and industrial arts programs. He lives in Round Rock, Texas.

Glenn E. Baker is Professor Emeritus of Industrial Technology at Texas A&M University in College Station, Texas. He is the author of a number of books, including *Carpentry & Construction,* Fourth Edition. He lives in College Station, Texas.